James Marcy

The sciopticon manual

Explaining Marcy's new magic lantern and light, including magic lantern optics. Fifth

Edition

James Marcy

The sciopticon manual
Explaining Marcy's new magic lantern and light, including magic lantern optics. Fifth Edition

ISBN/EAN: 9783337270957

Printed in Europe, USA, Canada, Australia, Japan

Cover: Foto ©berggeist007 / pixelio.de

More available books at **www.hansebooks.com**

THE

SCIOPTICON MANUAL.

EXPLAINING

MARCY'S NEW MAGIC LANTERN,

AND LIGHT,

INCLUDING

MAGIC LANTERN OPTICS, EXPERIMENTS, PHOTOGRAPHING
AND COLORING SLIDES, ETC.

BY L. J. MARCY, OPTICIAN,
No. 1340 CHESTNUT STREET, PHILADELPHIA, PENNA.

FIFTH EDITION.

SHERMAN & CO., PRINTERS.
1874.

PREFACE.

VISIBLE illustrations are so highly esteemed among educationalists of the present day, that the announcement that *a greatly improved form of Magic Lantern has made its appearance* is very favorably received.

Between a desire for such an instrument, and the lack of definite information concerning it, many very naturally manifest both their interest and their caution by searching letters of inquiry. The inventor has endeavored, in hasty epistles, to state all the possibilities and impossibilities of the Sciopticon, and of lights, lenses, lantern slides, and tanks; but he finds it increasingly difficult to give each correspondent, individually, a full philosophical exposition. For such, therefore as desire a more detailed account than can be given in a hasty letter, or than is contained in the circular, which it is always a pleasure to forward on application, a Sciopticon Manual has been prepared, to serve as a guide-book to the various interesting portions of the Magic Lantern field in general, as well as to the Sciopticon in particular.

It treats of the Optical Image—without a lens, with a lens, in the camera obscura, in the eye, in the photographic camera, and on the screen; of the peculiarities of lenses, and the corrections required by lenses; of the peculiarities of the Sciopticon, and its construction and management; of dissolving views, phantasmagoria, and the ghost; of lantern slides in all their variety; of photographing slides by the wet-plate process, by the dry-plate process, by Marcy's Photographic Printing Apparatus, by the Sciopticon, and other processes; of how to paint slides, and of how to perform chemical experiments, &c.; to which is appended a catalogue, arranged to assist purchasers in making satisfactory selections.

Thus this Manual may take the place of private correspondence to a considerable extent, allowing in letters more space for business, and for an interchange of new ideas, with a view of making the Manual in subsequent editions more interesting, and the Sciopticon more useful.

It was at first my intention to give space to the subject of chemical lights, but reflecting that in this direction there is no lack of printed matter, and that the interest felt in the Sciopticon is owing mainly to its giving

good results with little trouble, I have concluded to *omit
the gas*, which would increase the bulk of the Manual,
without a corresponding addition to its usefulness.

PREFACE TO FIFTH EDITION.

THE lime light, in an improved form, having been introduced into the Sciopticon, it has become expedient to
append to the Sciopticon Manual, a description of the
apparatus and directions for its use.

The demand for Lantern projections is steadily on the
increase. A fine photograph (and what can be finer?)
projected upon a large screen, before a thousand spectators, gives, it is safe to say, ten thousand times the
satisfaction that one alone with his stereoscope receives
from it. The appreciation is cumulative. "The more
the merrier," is the philosophy of it.

The Sciopticon with its oil lamp, rather than with its
lime light, continues to be the choice of the many, because its use is convenient and inexpensive. There are
purposes and occasions however for which the lime light
is a necessity. The *gas* therefore has now received its
full share of attention. Much of the added matter is
intended to assist those who have a Sciopticon, to provide themselves with interesting objects for exhibition,
without resort to a large assortment of expensive slides.

CONTENTS OF MANUAL.

(vii)

CONTENTS OF CATALOGUE.

INTRODUCTION.

THE SCIOPTICON (pronounced Si-op-ti-con), is by far
tthe most convenient and easily managed of any form of
Magic Lantern. Its ridge of wide, intensified double
fflame, lying lengthwise in the axis of the condensing
llenses, gives it much greater efficiency than any other
llamp-illuminated lantern.

All who have become acquainted with this new in-
sstrument, see in it the accomplishment of what has long
lbeen greatly desired by those who appreciate the value
cof visible illustrations as a means of imparting instruc-
ttion and of affording rational amusement.

Confessedly, the medieval magicians with their *lan-
iterne magique* effected little good by their incantations
:and ghostly spectres. But modern educators have
lhigher aims and better means at hand. Their lenses
:are greatly improved in form and quality. The pho-
ttographer secures images of all that is interesting or

beautiful in nature and art. Literature and the sciences
teem with pictorial illustrations, from which choice selections can be easily copied for lantern slides. And
now the Sciopticon, with its own peculiar light for all
ordinary occasions, and with the oxy-hydrogen light for
occasions extraordinary, comes in to show up what is
thus made ready.

In form and construction the Sciopticon is very unlike
that relic of the middle ages, the old magic lantern.
Those who are interested in the philosophy involved in
it, in the peculiarities pertaining to it, in the practical
management of it, in making and selecting slides for it,
in performing scientific experiments with it, and in pro-
moting the interest of education by it—will do well to
inquire within.

SCIOPTICON MANUAL.

CHAPTER I.

THE CAMERA OBSCURA.

A picture formed by rays of light from the several parts of an object as seen at *A* (Fig. 1), is called an image; and the *chamber* in which it is formed, and from which all light is excluded, except what enters a small hole as at *S*, is called a *camera obscura*.

This dark chamber claims attention here

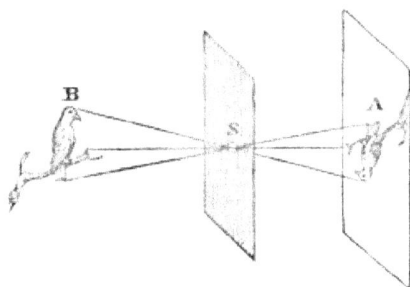

Fig. 1.

because its properties, which are common to the eye, the Sciopticon, and all forms of the camera, are seen on a broad scale, and may be readily verified by inexpensive experiments. To avoid confusion the pencils of light are represented in Fig. 1 without marginal rays needing to be focused by a convex lens. Explaining the following properties in connection with a diagram without

2

a lens, may help to correct the common impression that these properties and a convex lens are inseparable.

THE INVERTED IMAGE.

Of the rays of light proceeding in all directions from every point in the object *B*, as from all illuminated objects, just an assortment, that is to say a pencil from each point will enter the hole at *S*, just in line to fall in reverse order upon the screen *A*, forming an inverted image.

The picture results from admitting just an assortment and excluding all the rest.

COLOR AND SHADING.

Light from each of the several parts of the object *B* illuminates with its peculiar color and relative intensity each corresponding part of the image, so that it is seen in natural light and shade, and in natural colors. The photographer can fix the relative shading, but he cannot, as yet, fix the colors.

MOTION.

If, for example, the bird moves to a higher perch, the pencils of light will fall to a lower place on the screen, and so any movement of the object which alters the direction of the pencils of light, will give a reverse movement to the image.

SIZE.

By inspecting the angle of extreme rays it will be seen that the image in this case is smaller than the object, because it is nearer the aperture; so in all cases, the relative size of the image depends on its relative distance from the aperture.

SIZE ON THE RETINA.

A A (Fig. 2) represents the visible portion of the sclerotica of the human eye, which is a camera obscura in perfection. *I I* represents the *iris* (or window shutter), in the middle of which is the pupil (or aperture). As the retina is only about half an inch behind the optical centre, it follows that the images of distant objects upon it must be very minute.

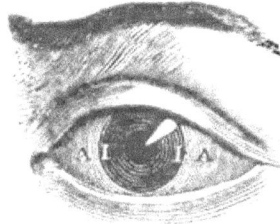

Fig. 2.

For example, the figure of a man 6 feet high, seen at a distance of 40 feet, produces an image upon the retina the height of which is about $\frac{1}{14}$th part of an inch. The face of such an image is included in a circle whose diameter is about $\frac{1}{12}$th of the height, and therefore occupies on the retina a circle whose diameter is about the $\frac{1}{170}$th part of an inch; nevertheless within this circle, the eyes, nose, and lineaments are distinctly seen. The diameter of the eye is about $\frac{1}{12}$th of that of the face, and therefore, though distinctly seen, does not occupy upon the retina a space exceeding $\frac{1}{1000000}$th of a square inch. How infinitely delicate must be the structure of the retina or canvas on which this exquisite miniature is delineated to receive and transmit details so minute with such marvellous precision !

SIZE IN THE PHOTOGRAPHIC CAMERA.

A man 6 feet high, standing for his picture 10 feet from a camera tube whose lenses require the screen of ground-glass adjusted to 5 inches, gives a 3 inch picture. For we have (in inches)

$$120 : 5 : : 72 : x \text{ hence } x = \tfrac{5 \times 72}{120} = 3.$$

MAGNIFIED IMAGE.

This rule works both ways; for suppose a screen to be suspended in place of the man, the space being darkened, and suppose the three-inch inverted image to be "fixed" and highly illuminated, then a secondary image will be projected upon the screen ten feet from the lens, corresponding to the original object in size and position.

INFERENCE 1.

A good portrait objective for the camera is also suitable for a lantern objective; for the lines of light and the angles are in both cases the same.

INFERENCE 2.

The light, if reflected from the three-inch picture, radiates so as to cover 100 times as much surface on the magnified image. Now, as a very small fraction of this reflected light is re-reflected to the eye of the observer, it seems a hopeless undertaking, to make the opaque lantern practically useful in showing the images of small paper photographs, on a large scale, with any ordinary flames, however well arranged.

INFERENCE 3.

With an intense light at a point behind the three-inch transparency, converged by a condenser, so as to enter the objective through all points of the picture, the magnified image is illuminated with incident rays concentrated, and its exhibition becomes a success.

INFERENCE 4.

Additional light outside this point (as some recommend), would not fall in line with the objective so as to improve the illumination; while the additional heat and diffused light would be very objectionable.

EXPERIMENTAL VERIFICATION.

These properties of the camera obscura, thus far considered, may receive more lively illustrations by actually darkening a room and admitting light through, say an inch hole. A room with but one window, and that looking from the sun, and towards objects illuminated by sunlight, is to be preferred. A lens, if one is used, of long focal distance (nearly flat) gives more room for spectators before the screen. The images, if the lens has short focus, may be better seen on the back of a semi-transparent screen by transmitted light, as they are seen on the ground-glass in a photographic camera. These moving pictures of busy life and wavy trees, of curling smoke and floating clouds, are peculiarly pleasing and beautiful, as well as suggestive of important principles in optics.

INDISTINCTNESS.

Fig. 1 fails of showing the divergence of each pencil of light to the size of the aperture as seen at c (Fig. 3);

Fig. 3.

a property which renders the image indistinct, from the consequent overlapping of the blunt ends, so to speak, of innumerable pencils.

CONVEX LENS.

In accordance with the law of refraction, rays as from d (Fig. 4) are bent towards a perpendicular in entering the convex lens l, and from a perpendicular in emerging from it. As the result of these refractions they meet at *f*. The converging power of lenses, of uniform substance and density, is in proportion to their

Fig. 4.

degree of convexity. For a lens to sharpen the image, the screen must be *adjusted* to the focal distance.

STOPS.

To get on the same plate something like distinct images of objects at various distances, a " stop " is used by the photographer, which, though it necessitates long exposure, secures "depth of focus." This expedient of having a small aperture is also resorted to for lessening the defects or aberrations of lenses, just as the aperture *b* (Fig. 3) is made small to lessen the greater defect of having no lens.

Stops are not used in the Sciopticon objective, because all portions of the picture-slide are in the same plane, and because sharpness produced by stops is always at the expense of light.

CHAPTER II.

THE CORRECTIONS REQUIRED BY LENSES.

The corrections required by lenses (as well as everything photographic), is well set forth in Dr. Vogel's Handbook of Photography. The use of such diagrams, as are here appropriated, is kindly allowed by the American publishers, Benerman & Wilson. Of course one may successfully operate the Sciopticon, or even excel in photography, without a critical knowledge of lenses; but a very short, connected showing of their properties, with diagrams, will doubtless prove acceptable to many who use the Sciopticon, or who are interested in photography.

THE FORM OF LENSES.

The convex, or converging lenses. 1, 2, and 3 (Fig. 5), called biconvex, plano-convex, and meniscus, are thicker

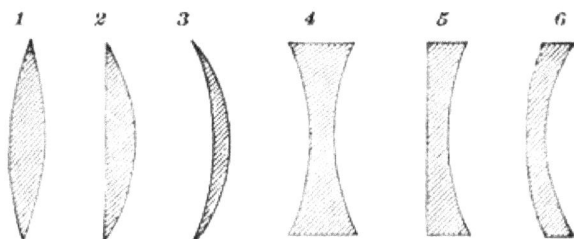

Fig. 5.

in the centre than on the margin. The concave, or dispersing lenses, 4, 5, and 6, called biconcave, plano-concave, and concavo-convex, are thinner in the centre than on the margin. A line through the centre of these lenses, from side to side, would show the axis of each lens.

PENCILS OF RAYS AND THEIR ILLUSTRATIONS.

A pencil of rays considered in reference to its direction and the points in the image which it illuminates, may be represented by a simple straight line, as in Fig. 1; but in most cases, when the action of lenses on its rays is considered, it must be shown as a bundle of rays, as in Fig. 4. The pencil in Fig. 6 differs from df in Fig. 4, in having middle rays represented as well as marginal, and

in having them proceed from a point too distant to be
shown. The rays of a pencil from a point 100 times
further from the lens than is the image, are about parallel, and their focus is called the focus of parallel rays, or
principal focus. A real pencil is composed of innumerable
rays, and such pencils from innumerable points in the
object meet and cross at the lens on their way to corresponding points in the image, and wonderful to tell,
no one is switched from the track for another, and there
are no collisions. An explanation of one answers for
countless millions.

SPHERICAL ABERRATION.

It is seen (Fig. 6) that the marginal rays d d must be
more refracted, or bent, than the more central rays f f,

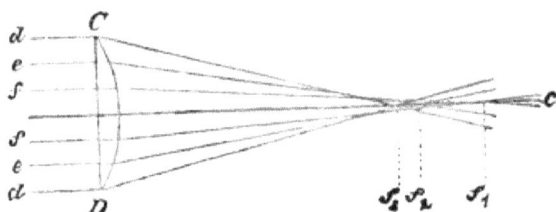

Fig. 6.

in order to meet the axial rays at f_1, and so it is seen
that the margin of the lens C D has a greater refracting
angle than the more central portions. But the trouble
is, the refracting at the margin is overdone, so that the
rays d d meet the axial ray at f_3 instead of at f_1. Hence
if a ground-glass has been placed at f_1, the marginal rays
which have intersected the axis at f_3 will form a circle
of dispersion about f_1. The diameter of this circle is
called the lateral aberration, and the distance between
f_3 and f_1 is called the longitudinal aberration. As a con-

sequence of this want of coincidence between the foci of the central and marginal rays the picture on the screen, or ground-glass, will appear blurred and ill defined.

We can conceive of a lens with a gradually lessening degree of convexity towards the margin, causing the foci to coincide, but lenses cannot well be ground in this form. The crystalline lens in the eye is supposed to cause the foci to coincide by an increase of density towards its centre, but such an arrangement of matter would be impracticable in art. Much is gained by reversing the lens, for spherical aberration is four times as great when the parallel rays enter its plane surface, as when they enter its convex surface.

Much is gained by a combination of lenses so that the refracting angle may be less in each. Were the marginal rays *d d* cut off by a stop, the aberration would be less, as we can see by tracing them in the diagram, but the illumination would also be less by so much.

DISTORTION.

When we focus with a single lens with a front stop

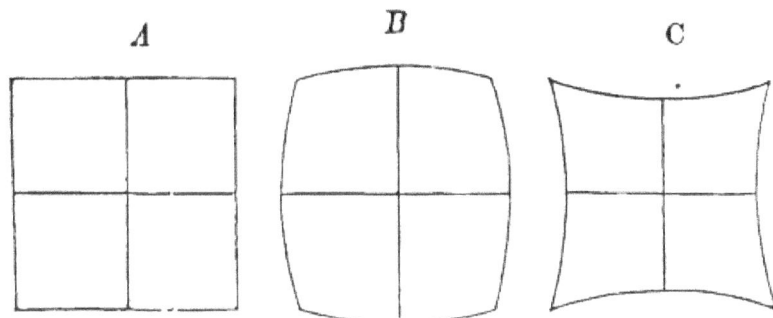

Fig. **7.**

sharply on a square, *A* (Fig. 7) the resulting picture will not appear square, but barrel-shaped, as at *B*. When we substitute a lens with the stop in the rear, the curves will be reversed, as at *C.* This property is based on the fact that the marginal rays of the field of view strike the lens under a larger angle than the central rays, and consequently suffer a greater refraction.

Of the simple form of lenses, the meniscus, with its concave side to the object, shows it the least. But it is best overcome by a combination of lenses with central stops.

CURVED FIELD.

This error is not caused by spherical aberration, for it occurs with all perfectly aplanatic lenses, but by the curve of the image, as is shown by the arrow, Fig. 8.

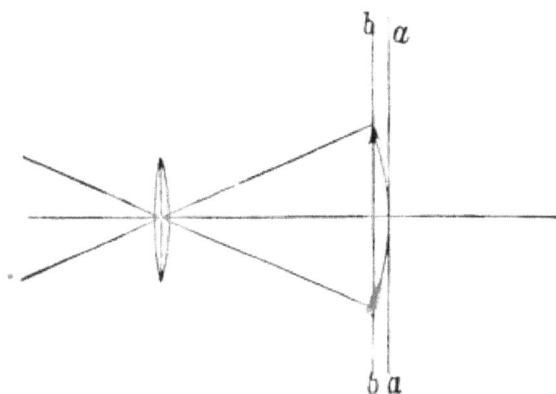

Fig. 8.

When the ground-glass is placed at *a a*, only the central part will appear sharp; when at *b b*, only the points of the arrow will appear well defined. This error is avoided by combination of lenses with suitable curves, and by stops.

CHROMATIC ABERRATION, OR DISPERSION OF COLOR.

White light is separated by a prism into the seven primary colors; violet, indigo, blue, green, yellow, orange, red.

As a lens is analogous to a system of prisms, and as violet is more refrangible than red, the violet rays $v\ v$ (Fig. 9) will intersect the axis closer to the lens than the red rays $r\ r$. This error is corrected by combining a concave lens of flint-glass with a convex lens of crown-glass, so as to neutralize their contrary dispersions.

The concave flint-glass lens f (Fig. 12), which has great dispersive power in proportion to its curves, diverges the violet more than the red, while the convex crown-glass lens converges the violet more than the red, so we have in both combined an achromatic convex lens. As the chemical rays are in the violet end of the spectrum, the photographer may succeed with an im-

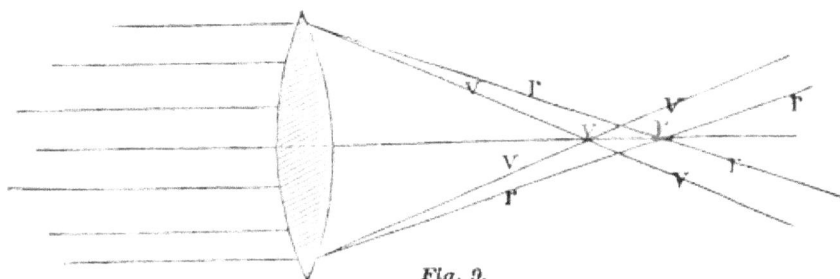

Fig. 9.

perfectly corrected lens by having the sensitive plate a little nearer the lens than the focus of luminous rays on the ground-glass would indicate. Lenses without chromatic aberration are called achromatic. The term aplanatic means without wandering, and may apply to lenses corrected of both spherical and chromatic aberration.

UNEQUAL ILLUMINATION.

We can see that the brightness of an image decreases from the centre towards the margin. The diameter of the pencil g g passing through the lens parallel to its axis, is of the same size as the opening of the stop B, and exceeds the diameter of the oblique bundle of rays. Besides, the oblique rays lose considerable light by reflection, which may in part be re-reflected upon the image,

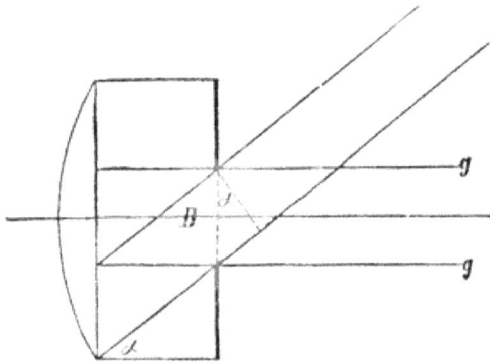

Fig. 10.

detracting from its distinctness. With a short exposure in the camera, this unequal illumination causes an under-exposure at the margin. In the Sciopticon it is even exaggerated by the reflector, but we generally wish the objects occupying the central portion of the *"field of view"* to stand out more clearly in the illuminated disk.

CORRECTIONS IN THE EYE.

Spherical aberration and distortion in the eye are corrected (it is believed) by the greater density of the crystalline lens about its centre than towards the margin, where the refracting angle would otherwise be too great.

Chromatic aberration is corrected by the combined

action of the crystalline lens and the vitreous and aqueous humors. As the retina, $R R R$ (Fig. 11), is concave, the centre of concavity being the optical centre, there is no need of adjusting the focus to a flat field.

The eye is readily, for the most part unconsciously, adjusted, so that an object upon which we fix our attention is at once in the centre of the field of view, and is focused according to its distance.

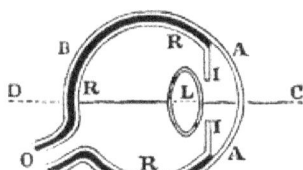

Fig. 11.

These five troublesome properties enumerated in this chapter, are thus, in the eye, harmoniously reconciled. In art we lack the peculiar crystalline lens, and the concavity of field. Making amends for this lack interferes with other corrections. Efforts of various makers to effect the best compromise for particular kinds of work has given rise to lenses, in variety too numerous here to particularize.

THE LANDSCAPE LENS.

This simple achromatic lens (Fig. 12) is the oldest photographic lens in existence. It is composed of the

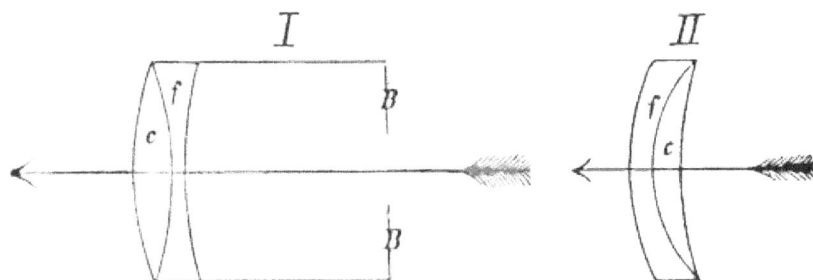

Fig. 12.

concave lens of flint-glass f, and the convex lens of crown-glass c.

Among the modified forms, the Dallmeyer Landscape Lens, which consists of three lenses cemented together, a central one of flint-glass and two outer ones of different kinds of crown-glass, gives better results. The stop $B\,B$ is generally one-fifth of the focal length distant from the lens, and consequently cuts off much of the light. In the earlier days of photography a person had to sit in front of such a lens, in a strong light, for several minutes. That in this way no artistically perfect pictures could be made is self-evident, and so it became necessary for portrait photographers to have a lens that would work satisfactorily with a larger opening.

THE PORTRAIT OBJECTIVE.

This invention is no accident, but the result of a thorough theoretical calculation. It is a double objective with two unequal lenses, with or without central stops between.

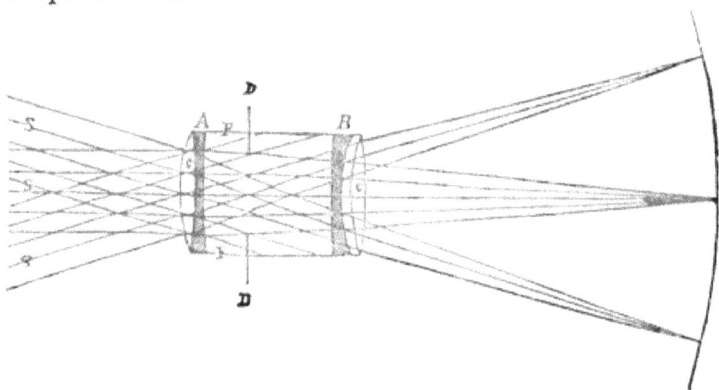

Fig. 13.

The front lens A consists of a biconvex crown, and an almost plano-concave flint-glass lens, cemented together with Canada balsam. The back lens B consists of an almost plano-concave flint, and a biconvex crown-glass lens, separated by a ring. It is seen that some marginal rays entering the lens *A* do not reach the lens *B*, because of the length of the tube *F F*, which effects about the same result as the slight stop *D D*.

The curvature of the field is somewhat exaggerated in the diagram, to remind us that an aplanatic lens cannot give a perfectly flat field without a stop. In this general form all portrait lenses mostly coincide, differing in regard to the focal length of the separate objectives *A* and *B*, the distance and size of the same, the position of the stops, &c.

Considering Fig. 13 as representing a portrait objective, the pencils *S S S* proceed from an object comparatively large and distant, forming a small image in the camera. Considering the diagram as representing a lantern objective, the order is reversed. A small transparency is in place of the curved line, which in its turn becomes the focus of incident rays, projecting upon a screen a comparatively large and distant image where the pencils *S S S*, if extended, meet in a focus of refraction. *B* is called the back lens in either case, as it is back next to the instrument to which the tube is attached.

CHAPTER III.

THE CONSTRUCTION OF THE SCIOPTICON.

THE PORTRAIT OBJECTIVE.—This objective (Fig. 13 or 15) is made for the camera, and is known mostly in its relations to photography. An objective, however, that with large opening, will give proper direction to rays from a large object to a small image in the camera, will answer equally well in giving direction to rays from the small picture in the Sciopticon back to life-size on a screen; both object and image being in the conjugate foci in either case.

THE PLAIN LANTERN OBJECTIVE.—This objective, like the achromatic portrait objective (Fig. 13 or 15), has the advantage of a front and a back lens, A B (Fig. 14), so far apart that the tube serves as a stop for marginal rays with comparatively small loss of light. The front lens A is a meniscus of crown-glass, whose tube slides into a larger tube which holds the plano-convex crown-glass lens B. Arranged as in the diagram, the effect is scarcely inferior, so far as common observation goes, to that of the most expensive combinations. With the front tube reversed, so as to bring A near to B, the image is larger but less distinct. With only one lens the image is smaller. These different arrangements give the three powers commonly attributed to lenses mounted in this form.

NO LOSS OF LIGHT FROM USING AN OBJECTIVE OF LOW POWER WITH A CONCENTRATED LIGHT.

Were the picture p (Fig. 14) made luminous by light shining upon it, as in an opaque lantern, the light would radiate in all directions, and in accordance with the law of radiation would lose intensity with distance, and so a lens of short focus, being nearer, would bring more light

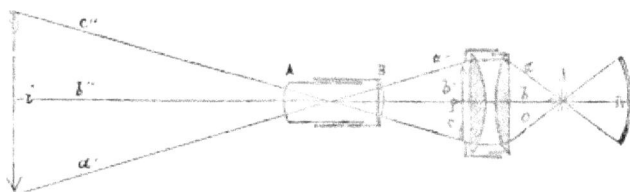

Fig. 14.

to bear on the image. But the light in fact proceeds from *l*, and what passes through the condenser and picture becomes a cone of light, with its apex in the objective.

This cone of light must be regulated by the position or power of the condenser so as to fall within the compass of the objective. Except for some reflections from the surfaces of the glasses, the light would not illuminate the room at all in its passage, and there would not be even the small need there is of the hood *i* (Fig. 15).

ADVANTAGES OF A CONCENTRATED LIGHT.

Could the light proceed from a mathematical point behind a faultless condenser, the pencils *a′ b′ c′* (Fig. 14) would be without marginal rays, and there would be nothing for the objective to do, for its office is to bring the marginal rays of each pencil to coincide with the axial ray. Were we to adjust the aperture of our window shutter (Fig. 1), to the apex of the cone, there would be no rays for the shutter to cut off. The image would be formed anywhere within reach of the light, either with or without lens or stop. But in reality even the most concentrated light occupies some space with countless radiant points, so near together however that

3

their radiations cross at a very small angle. The rays which meet and cross at each point in the picture p, become a pencil of diverging rays to be focused on the screen by the objective.

We conclude, therefore, that the more concentrated the light, the nearer in each pencil will the marginal rays coincide with the axial ray, and the less will the imperfections of lenses become manifest.

THE CONDENSER.

The condenser is formed of combined lenses, because the refracting angles would be too great in a single lens of sufficient diameter and short focal distance.

The simplest arrangement is where two plano-convex lenses are combined, with their curved surfaces inwards, as in Fig. 14, or at $p\,q$ in Fig. 15. The shorter the focus of the condenser, the shorter, with a given objective, must be its distance from the light; it will collect more light, but it will be in more danger of breakage from the heat. In the Sciopticon the space between the lens q (Fig. 15) and the front of the flame is only about two and a half inches, but the glass G, and the air between it and q rising up and out at A, makes it perfectly secure. The condensing lenses in the Sciopticon are usually each 4 inches in diameter; but a 4½ inch front condensing lens, p, is used to advantage for slides larger than the standard size, and to show fully the corners of some of the ordinary square transparencies.

THE REFLECTOR.

The reflector r (Fig. 14) is so adjusted that the light is in the centre of concavity. Each ray is reflected back to its starting-point, and continues in line with the incident rays *a b c*. Some advocate a larger reflector, a parabolic reflector, or reflectors at the sides; but unless the reflected light coincides with *a b c*, or nearly so, it is not transmitted by the objective lens.

With proper adjustments, the light *l*, with that from the reflector, is focused at the objective tube, of a comparative size proportionate to its relative distance from the condenser; and the picture *p* is focused upon the screen at *i* of a comparative size proportionate to its relative distance from the objective. As represented in Fig. 14, the smallest diameter of the cone of light in the objective *A B* would be twice that of the point of light *l*, and the height of the image *i* would be twice that of the picture *p*.

To project a picture to a great distance without too much enlargement, the objective must be of low power and carried forward; and the light should be from a point (as in the calcium light) to avoid loss, and should be carefully adjusted to secure even illumination.

The longitudinal ridge of light *E* (Fig. 15), with a medium objective gives uniformly good results without perplexing experimental adjustments.

VARIOUS MODES OF LANTERN ILLUMINATION.

The Hydro-Oxy-Calcium light, or lime made incandescent by a jet of hydrogen and oxygen in flame upon it, is the most brilliant available light. Its concentrated form adds greatly to its value for the lantern. The Oxy-Calcium light produced by a jet of oxygen on to lime, through an alcohol flame, is much inferior to what is produced by the mixed gases. It is much used, because one gas is easier to manage than two, and it is counted safer. A lantern with a calcium light of either kind is commonly called a *Stereopticon*.

The Magnesium light is excellent in photography, but it gives off troublesome fumes, and for a continued lantern exhibition is too liable, even with well-regulated clockwork, to leave the lookers-on in sudden darkness.

The Electric light is intense and concentrated, but it requires too much apparatus to be available.

The above so-called chemical lights, are, if *well* managed, much brighter than flame, even at its best. The Sciopticon has a double flame, which is not only very bright, but gives much more distinctness to the image, by its standing edgewise to the condenser instead of broadside, as a single flame must, to prevent its casting a dark shadow on the disk. No lantern of any sort can compare with the Sciopticon in point of convenience. We may conclude that the Hydro-Oxy-Calcium light is best for exhibitions on a large scale, and as for the rest, the Sciopticon is desirable as combining efficiency and great convenience.

THE SIZE PROPER FOR THE ILLUMINATED DISK.

The image enlarges in area, and diminishes proportionately in brightness, as the distance of the lantern from the screen increases. A disk of six or seven feet is about right for figures, statuary, &c., to give brightness and not an unnatural size; while landscapes, &c., appear better on a disk of eight or ten feet, or more. With an objective of about four inches back focus, as is most used on the Sciopticon, a distance of about sixteen feet from the screen gives a disk of about eight feet. The arrangement can be varied to suit circumstances.

ANALYSIS OF THE SCIOPTICON.

The lenses, mountings, &c., Fig. 15, are shown in section. What is left of the frame and cylinder, the lamp, chimney, reflector, &c., are shown in perspective. The parts are as follows:

a b—Front combination of the objective cemented together.

c d—Back combination separated by a ring. If the cells holding these combinations are unscrewed and the lenses removed, they must be returned in the same order and position as seen in the diagram. There is no need of removing them. Even the outer surfaces of *a* and *d* will seldom need dusting if kept in a clean place with the caps closed. They should not be fingered, and the brush or fabric used for dusting them should be clean and soft.

e—Milled head for adjusting the focus.

f f—Flange attached to the projecting wooden ring *g g*.

The tube here represented is a quarter-size portrait camera tube of 4¼ inches back focus, requiring an aperture in *g h* of 2⅜ inches in diameter. If a larger tube is used, the aperture in *g h* has to be enlarged. If the back focus is more than 5 inches, the extension front *h k* must be drawn out more or less from the main body, as is shown in the diagram. If the focus is shorter than 3½ inches, the ring *g g* is removed, letting the flange *f* back to *h*.

h h h'—Wooden frame of the extension front; *h'* sliding in a groove within the body-frame *l l*.

i—Top of the hood covering light dispersed by reflection. The near side is cut away to show the screen *k*; the edge of the remaining side is seen beyond *k*.

k—Is now modified into a horizontal lid, which shuts up over the lens *d*, darkening the picture on the screen like a falling curtain.

l l—Portion of the wooden frame, the rest being mostly cut away to show the lamp, and how the extension front slides in its groove.

m—Claw attached to the front foot.

n—Flange under the back foot. On the top of the Sciopticon case, or box in which it is carried, and which can be placed upon a stand or table to elevate the instrument to proper height while in use, are two round-headed screws, slightly raised, and at the distance apart of *m n*; *m* clings to one, and *n* slips under the other,

thus holding the instrument firmly in place. When a pair is used
for dissolving views, the fronts are thus held in a fixed position,
while the rear ends may be spread apart till the disks on the screen
coincide.

o o'—Stage and spring for wooden-mounted pictures. The operator standing behind, slides a picture horizontally in at *o*, letting it
bear against the condenser mounting, and letting it project equally

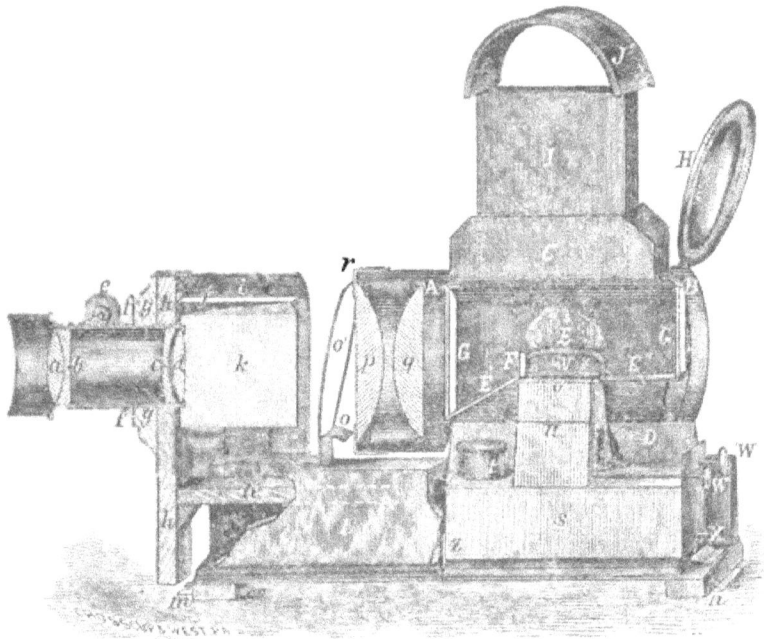

Fig. 15.

both sides of the cylinder. The picture is drawn out with the left
hand, while with the right hand another is made to follow in its
place, so as not to show the white disk on the screen.

p q—Condensing lenses. Lenses when taken from a damp or
cold place are apt to become covered with moisture, which shades
the pictures. It is better when this is likely to be the case, to let
the instrument stand in a warm room awhile, or else to draw the
lenses apart and dry them before beginning an exhibition.

r—Brass ring, holding the condenser cells suspended in the cylin-

der, so as not to be anywhere in contact with it. The ring shuts
over the end of the cylinder like the cover of a tin pail. To remove the condenser, the extension front is drawn off, and the stage
o is lifted out of its place.

s—Lamp cup for kerosene oil. It holds three gills, or enough to
last about 4 hours. When it has to be moved about much, it is
better not to fill it more than two-thirds full, for if any oil gets outside, it gives off its offensive smell; while if there is no oil outside
there is no smell from it in the least. When packed for transportation, the oil should be thoroughly drained off.

t—Nozzle to admit the oil. It is large, so that if a wick is carelessly turned down into the cup, it can be fished out with a bent
wire.

u—Side of one of the two tubes, showing how the conduction of
heat downwards is counteracted by breaking the connection in the
metal. It is made of tin, for the reason that it is a slower conductor
than brass.

v v—Tops of the two tubes. They carry No. 3 wicks, which are
an inch and a half wide. The lamp being taken out, the wicks are
pushed down the tubes till they are caught by the ratchet-wheels
and drawn down. Should a loose thread of the wick get clogged
in the wheels it must be drawn out and cut off. The ratchet-wheels
could be made to bear tighter on the wicks by pounding gently
along the bottom of the tubes, but such a necessity is not likely to
happen.

w w—Buttons for adjusting the wicks; both are turned *inward*
to raise the wicks, and *outward* to draw them down.

x—Spring for holding the lamp.

z—Stop, preventing the lamp from sliding in too far.

A B—Portions of the cylinder not cut away, seen beyond the
condenser and flame-chamber.

C—Portion of the cylinder turned up, to give free ventilation all
about the flame-chamber.

D—Portion of the cylinder turned down and supported by the
wooden frame.

E E' E''—Bottom of the flame-chamber. It is not supported by
contact with the lamp, thus avoiding the conduction of heat downwards. The slot through which the flame ascends is two inches
long by half an inch wide. E' answers to the deflecting cap of **a**

common lamp. E'' is level, to allow the lamp (*the wicks being turned down*) to slide in and out. E slopes so as not to shade the light from the condenser.

F—Narrow strip of glass, quarter of an inch wide, held in a socket before the flame, to give upward direction to heated air. . It will not crack from heat because it is so narrow, and without obstructing light it takes from the glass G its liability to crack.

G—Front of flame-chamber glass. It is now held in a tin frame by a wire ring, so that should it crack, it is still kept in place without harming the effect on the screen.

G'—Back flame-chamber glass. The lamp is lighted by removing this glass, and reaching the wicks with a lighted match. G G' must be in place to secure the draft. F, especially since the introduction of the tin frame for G, is scarcely necessary.

H—Reflector, used also to close the rear of the cylinder. The centre of concavity is at E', so that reflected rays are thus made to coincide with incident rays from E' to the condenser.

I—Chimney, giving large outlet to heated air.

J—Chimney cap, for darkening the outlet. It may be raised to increase the draft, when the lamp gets to burning freely enough to bear it.

PACKING.

No instrument is forwarded without being first proved by careful trial. The oil is then poured off, and the lamp burned awhile afterwards, to prevent any further drainage should it be shipped wrong side up. Let this precaution be taken by all who pack the instrument for transportation, that there may be none of the offensive smell of oil when the instrument is unpacked and used.

The wicks are left in the tubes, ready for use. Four extra wicks, with the narrow glass, F, are tied together, which with a dozen flame-chamber glasses, G G', are sent with each instrument. F is removed, as it is liable to fall out if inverted. For the most part G G' are left in place. Packing is placed between G and q, to prevent their getting out of place and scratching against each other; also between G' and H, and between the condensing lenses.

The cap J is removed and placed behind the chimney. The whole is snugly packed in a box with stuffing, and the cover fastened on with screws. These particulars may be advantageously referred to in case of repacking by the purchaser or borrower.

RULES FOR OPERATING THE SCIOPTICON.

In unpacking a new instrument the parts must be separated, to remove the packing papers.

Dust them if necessary.

For the lenses and reflector use a duster that is soft and clean.

Warm and dry the condensing lenses if inclined to fog.

Adjust F, G, G', J, and the lenses, as seen in Fig. 15.

Shut the extension front back to its place.

Fill the lamp about two-thirds full with standard kerosene oil. The fire test should be 110° at least; that of Pratt's astral oil is 145°.

Avoid carelessly tilting the lamp when it is very full, and so avoid the smell of oil evaporating from the outside surface.

Turn down the wicks, so they will not rub against the deflecting plate while withdrawing or inserting the lamp.

It is convenient to stand the instrument so as to be about breast high.

The image enlarges as the distance of the instrument from the screen increases. With a medium objective, a distance of sixteen feet gives a disk of eight feet, &c.

Exhibitions of this sort appear to the best advantage

in the evening; shutting out daylight is not only troublesome, but the eye is not prepared for the contrast.
Lights should be turned down near the screen, but may
be left dimly burning in the distance, or out of range
of the screen.

Light the lamp in the instrument, as it stands in the
diagram, by removing the back glass, G', turning up the
wicks by a turn inward of the buttons $w\ w$, and reaching the wicks $V\ V$ through E with a lighted match.
To avoid smoke, turn the wicks almost down again till
the glass is replaced.

Turn up the flames evenly about half an inch at first;
they will rise a little after the wicks are warm, when
they may need looking to again, after which they will
stand steady without requiring further attention.

Put out the light by drawing the wicks down with a
turn of the buttons outward, and then blowing under
the reflector.

The wicks may be trimmed when the lamp is taken
out to be filled; cut them level; it may be done more
evenly by only removing the black part.

If kept in a dry place the reflector will keep its polish
for a long time; it is protected by a film which should
not be rubbed.

While exhibiting, the operator should stand behind
the instrument, having the slides arranged at his right,
in the proper order and inverted position required for
exhibition. If the instrument is in front of the screen,
the wire ring fastening the double glass into the wooden
mounts should be towards the condenser, in order to
show the views in a right-handed position. Some oper-
ators mark what should be the upper right hand corner
of each picture, with a piece of white paper, or a notch.

Pass the slides in with the right hand, level and true,

without jumping them about. The stage o slants down to the condenser, to keep the slides down close to it.

Take the slides out with the left hand as others are pushed into place, so as to leave none of the white disk visible, and put them in their box as before. A slide standing endwise between those which have been used and those which have not, will keep them apart.

As photographers are giving increasing attention to preparing slides, there is an increasing proportion in the market of the size of half a stereoscopic view, or 3¼ inches square, bound with narrow binding. For these a wooden stage 9 inches long is attached to o o', so that, without crowding a picture out at the end, its successor may be pushed into its place, by the finger following to where the cylinder and stage intersect; with the left hand at the button attached to the back stop we may: 1. Close stop. 2. Slide in the picture. 3. Uncover—so that in the time of counting three we have changed the scene without any visible movement. This, well managed, is better than dissolving views poorly managed.

Tanks for insects, fish, chemical experiments, &c., &c., slide into the stage as easily as pictures. The stage being open at the top, with no bulky lantern case to obstruct it, is peculiarly suited to all such operations.

A slender wire in the direction r o', answers the purpose of a long rod pointing upward on the screen to explain the representations.

The simplicity and completeness of the Sciopticon are more evident in practice than may seem while considering so wide a range of details and contingencies. The advantage of having an instrument so completely under one's hand is not only felt by the operator, but the smoothness it gives to the exhibition is appreciated by spectators.

Beginners who wish to understand and operate the Sciopticon by explanations and directions which can be seen at a glance, may examine in connection with Fig. 15, the following

RECAPITULATION.

The front, *h h h'*, with its attachments, draws apart from the body of the instrument.   .

The stage *o o'* lifts out.

The condenser, *p q*, is drawn out by laying hold of the ring *r*.

The cells holding *p* and *q* draw apart.

The front flame-chamber glass *G* is held in place by the spring *A*, which can be reached through the opening over *A*.

With *h o p q G* removed, the narrow glass *F* (found packed with the extra wicks) is reached to position, and needs no further attention.

The portion of chimney attached to the cap *J*, telescopes into *I*.

The lamp *S* slides out horizontally, by raising the spring *X*.

With packing removed, glasses clean, lamp filled two-thirds full of standard kerosene oil, and all parts in place as seen in the cut, remove the back glass *G'*, and reach the wicks *v v* with a lighted match. Replace *G'*, and let the flames stand about one inch high.

See, specially, that an oil so inflammable as to light at the safety slit *u* is not used—that no oil is left outside the lamp-cup, to give off an offensive smell—that the wicks at *v v* are not raised to rub against the plate *F'* when the lamp slides in and out—that the flame-chamber glasses *G G'* are in place to secure draft, and that the oil is thoroughly drained out of the lamp-cup should the instrument have to be repacked for transportation by public conveyance.

Standing behind the instrument, placed about breast high—as upon its box on a stand or table—close down the reflector *H*, pass in the slides at *o o'* with the right hand, taking them out with the left as other slides take their places. Focus the picture by the milled head *e*, upon the screen, which may be distant sixteen feet, more or less, as it is desired to have the scenes on a larger or smaller scale.

k (unlike the cut) is horizontal, and turns up to give the appearance of a falling curtain on the screen.

THE SCREEN.

There can be nothing better for the projected pictures than the white-finished, whitewashed, or white-papered walls of many a lecture-room or dwelling. An appropriate space specially set apart and papered with white wall paper, having an outline, say of a wide recess or niche for statuary, is an inexpensive and not inelegant fixture, on which to display before the assembled household, without waste of room or trouble in arranging, the richest treasures of all the art galleries in Christendom. The time is coming, when for purposes of demonstration and illustration in the lecture-room, this *whiteboard* will rival the *blackboard*.

The best material in the market for a movable screen of good size, seems to be bleached sheeting of close texture, but not very fine, twelve-quarters wide. This gives us the material, nine feet square, for about two dollars. It has the advantage of being available whether the instrument is placed before or behind it. As, however, every pencil of light falling between the

open threads of the texture is lost, it is better, when
the instrument is invariably to be placed in front,
to cover the surface with whiting or paper, keeping
it smooth by mounting it on a roller. When illuminated from behind, the screen should be wet, to tighten
its texture and to make it translucent, and consequently
luminous on the side towards the spectators. It can be
wet and then stretched upon a frame, or first mounted
and then sprinkled to saturation. For home use, a sheet
may be stretched across the frame upon which the folding doors of most modern houses are hung, the doors
being thrown open at the commencement of the exhibition. A waxed screen is often recommended, but it is
little used on account of the difficulty of keeping it
smooth and clean. An unmounted screen can be quickly
put up in any room by procuring two strips of wood
about two inches square, and long enough to reach from
the floor to the ceiling; a side of the screen is tacked
to each one of these strips, which are then stretched
apart, and wedged up tightly between the floor and the
ceiling.

To widen the screen to more than nine feet, join the
added width to each side, rather than bring a seam into
the centre of the views.

A fine picture from within, upon oiled muslin, stretched
upon a frame, made to fit a show window, is always
greatly admired by all the passers-by. Such a framed
oiled screen, on a small scale, can also be conveniently
used in parlors, or in the doorway leading out from the
company.

Working behind the screen has in many cases decided
advantages, but the images can hardly be as bright by
transmitted light, and other things being equal, it is
better for the instrument to be in front.

DISSOLVING VIEWS.

THE STAND.—The peculiar stand represented in Fig. 16, is mostly the one used with Sciopticon dissolving apparatus, and so can better be described with it, but it is not necessarily a part of it.

It consists of a well-made walnut box, mounted on two pairs of adjustable legs, attached by fixed thumbscrews and nuts. The back legs are an inch or so shorter than those attached to the front at A, to elevate the range of the lanterns. The back of the stand may be known by the match-lighter G, and by its being necessary for the operator from behind to have the opening and the box of slides B at his right hand. The slide D stands on end, to separate the used from the unused slides.

When the apparatus is taken down, the legs swing together on their hinges, and are tied in a bundle; the open side of the box becomes the top; the instruments occupy the stalls E and F; the dissolver is drawn apart and placed alongside; the caps are removed from the chimney, and placed in the rear; the box of slides occupies the space in front; the swing shelf C becomes the lid and is locked down; the strap S and its mate, now hidden under the instruments, meet over the top for one carrier, or serve like the ears of a basket, for two.

But as a stand, as seen in the diagram, the front of the box becomes the baseboard, and like any other 13 by 17 inch board, affords suitable standing-room for the apparatus; it is more likely to keep it level than a separate board, as it is dovetailed and firmly fastened in place.

DISSOLVING APPARATUS ARRANGED.—The fronts of the sciopticons R and L, hold firmly by claws to two

Fig. 16.

screw-heads 7¼ inches apart; the flanges in the rear slide, under two similar screw-heads, holding the instruments down, but allowing them to spread till their disks coincide on the screen.

The construction of the dissolver is shown in Fig. 17, in its three parts. The crescent-shaped dissolver a is mounted on the arm b, as seen in Fig. 16, so as to cover alternately the tubes on R and L, as it swings from side to side. The horizontal part of b slips into c till the length of the united axle just allows the dissolver to swing clear of the tubes, and the whole is held in place by a socket-spring at each end of the baseboard.

The dissolver is operated by the handles at c, which are adjusted at the proper angle to limit the lateral movement of a to the distance between the tubes.

Light the lamps in their place by reaching the wicks with a lighted match, and attend to them at first to see that they burn steadily and evenly. Focus a picture in R, for example, while L is covered by the dissolver, and

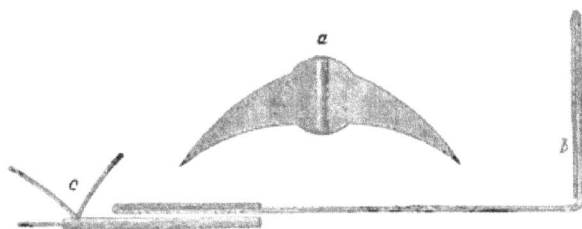

Fig. 17.

in L while R is covered; this reduces the disks to equal size on the screen. With the slides removed, and the dissolver in the position as shown in Fig. 16, spread the lanterns till the disks coincide.

DIRECTIONS FOR PRODUCING THE DISSOLVING EFFECTS. —With the lanterns lighted, and arranged as shown in Fig. 16, and a slide placed in each, then the gradual

4

moving of the dissolver will very mysteriously dissolve one view into another.

This effect is commonly produced with slides not specially arranged for the purpose, but it is desirable that they should be of similar size and shape, and that they should be put in evenly, so as to cover the same space on the screen.

Many slides are, however, selected and executed with special reference to their producing charming effects in dissolving.

They are mostly arranged in pairs, as some view in summer and the same in winter, by day and by night, interior and exterior, in sunshine and in storm, or humanity in opposite moods. Sometimes the series are more extended, as the Seasons, the Voyage of Life, &c., and sometimes they are in connection with chromatropes to represent volcanic action, conflagrations, fireworks, turning mills, &c. Suppose, for example, Saint Peter's, at Rome, is thrown upon the screen from R, and a night view of the same is placed in L; then as the dissolver is changed, Saint Peter's with its surroundings continues on the screen, but an appearance of night comes over it; the windows glitter with a thousand lights, and the moon makes its appearance in the heavens. Now, suppose a chromatrope, suited to the purpose, is placed in R, then as the change proceeds fireworks will rise from the darkness, and illumine the sky.

The snow effect is produced by a strip, usually of silk, with pin-holes all over its surface, mounted on rollers within a slide, so that when the silk is rolling up, snowflakes appear on the screen to be falling. Let, for example, a farm-house scene be projected from R upon the screen, amid all the glory of summer vegetation; place the snow slide in L, and let an assistant slowly roll it up while the dissolver passes over; the snow shows plainer and plainer, till nothing but the falling snow appears. Now place in *R* the same view in winter and turn back the dissolver; the storm subsides, and the farm-house scene again appears in the morning light, covered with the newly fallen snow of the winter's night.

To bring out statuary on a blue ground, a slide of blue glass, and usually one of red glass also is used. Change any scene, first into a red disk, then the red into blue, and then let a piece of statuary slowly come out into the blue ground, while the blue becomes darker and darker, till it ends in a blackness which seems to add vigor to the representation.

A beautiful effect is produced by a wheel chromatrope, used continuously in one of the lanterns, while a series is shown in the other, turning it inward and outward alternately, as the dissolving proceeds. It thus seems to suck up the vanishing scene as in a maelstrom, and to bring out its successor with scintillations of colored lights.

A pleasing effect is produced by showing a series of views in one lantern, and a veranda, or some appropriate design with opaque centre, with the other. If in adopting this suggestion, the veranda be focused for the edges of the field, and the view focused for the centre, a flat field is obtained over the entire disk. In this case, and in all cases when light from both lanterns is to appear, the dissolver is slipped up an inch higher, and kept in position as in Fig. 16.

The slow or dissolving process may become monotonous, and it is not always appropriate. We hardly like to see "Pilgrim" in his "Progress" fading away, while his *double* by his side is slowly growing in strength and vigor.   It is better to allow the axle of the dissolver to turn at once, flashing the change upon the disk.

Much use can be made of this expedient, as it is so easily effected in the apparatus represented. A duplicate picture placed in R and L in reverse order, the dissolver being changed back and forth with a sudden movement, will show an "about face" as of a person bowing to the company, a lion uneasy in his cage, &c.

Lightnings may thus be made to flash upon scenery, especially when the view is darkened somewhat by turning down its light a little, giving the appearance of a rising tempest.

Discretion and good taste should be observed in arranging the slides for an exhibition, so as not to mar beauty with caricature, or sacred scenes with what is ridiculous ; yet it is well to avoid monotony, for "variety is the spice of life."

Dissolving views, it must be confessed, are usually treated in a somewhat florid style by opticians, so it may be safe to make some abatement in anticipating the effects, especially of high-priced mechanical slides, lest when they chance to fall below the "Royal Polytechnic Institute in London," there should be a feeling of disappointment.

In the Sciopticon enterprise, it has been kept steadily in mind, to produce beautiful and useful results by the simplest means; and the desire is felt, not to make as large sales as possible, but to have every purchaser realize his highest expectations.

THE PHANTASMAGORIA.

To produce this effect, the operator should be on one side of the wet screen, and the spectators on the other.

Taking the instrument under his left arm, he should go up pretty close to the screen, and adjust the focus with his right hand; the image of course will be very small; he must then walk slowly backwards, at the same time adjusting the focus. As the image increases in size, it will appear to the spectators to be coming towards them; and then again let him walk up towards the screen, thus diminishing the image, and it will appear to them as if receding. The screen not being seen, the image appears to be suspended in the air, and the deception is complete, even to those accustomed to the exhibition. The focusing is most evenly and easily effected by prying the extension front out and in with the thumb and fingers of the right hand.

Slides producing the best phantasmagorial effect are those containing but one or two figures with a black background. In ancient times, the images from the phantasmagoria were thrown on the smoke arising from a chafing dish in which odors and drugs were burning, and by means of which many surprising and apparently supernatural effects were produced. As a relief from so closely following practical details, let us advert to the probable use made by ancient magicians, necromancers, and sorcerers, of these optical contrivances for producing supernatural illusions. In this we cannot do better than to quote from that eminent authority on optical science, Sir David Brewster:

" In the imperfect accounts which have reached us of these representations, we can trace all the elements of optical illusion. In the ancient temple of Hercules, at Tyre, Pliny mentions that there was a seat made of consecrated stone, 'from which the gods easily arose.' Esculapius often exhibited himself to his worshipers in the temple at Tarsus; and the Temple at Enguinum, in Sicily, was celebrated as the place where the goddesses exhibited themselves to mortals. Jambliches actually informs us that the ancient magicians caused the gods to appear among the vapors disengaged from fire.

"The character of these exhibitions in the ancient temple is so admirably depicted in the following passage of Damascius, quoted by M. Salverte, that we recognize all the optical effects which have been already described. 'In a manifestation,' says he, 'which ought not to be revealed, ... there appeared on the wall of the temple a mass of light, which at first seemed to be very remote; it transformed itself in coming nearer, into a face evidently divine and supernatural, of a severe aspect, but mixed with gentleness, and extremely beautiful. According to the institutions of a mysterious religion the Alexandrians honored it as Osiris and Adonis.'

"These and other allusions to the operations of the ancient magic, though sufficiently indicative of the methods which were employed, are too meagre to convey any idea of the splendid and imposing exhibitions which must have been displayed. A national system of deception, intended as an instrument of government, must have brought into requisition not merely the scientific skill of the age, but a variety of subsidiary contrivances, calculated to astonish the beholder, to confound his judgment, to dazzle his senses, and to give a predominant influence to the peculiar imposture which it was thought desirable to establish. The grandeur of the means may be inferred from their efficacy, and from the extent of their influence.

"This defect, however, is to a certain degree supplied by an account of a modern necromancy, which has been left us by the celebrated Benvenuto Cellini, and in which he himself performed an active part.

"'It happened,' says he, 'through a variety of odd accidents, that I made acquaintance with a Sicilian priest, who was a man of genius, and well versed in the Latin and Greek authors. Happening one day to have some conversation with him when the subject turned upon the art of necromancy, I, who had a great desire to know something of the matter, told him, that I had all my life felt a curiosity to be acquainted with the mysteries of this art.

"'The priest made answer, "that the man must be of a resolute and steady temper who enters upon that study." I replied, "that I had fortitude and resolution enough, if I could but find an oppor-

tunity." The priest subjoined, "If you think you have the heart
to venture, I will give you all the satisfaction you can desire."
Thus we agreed to enter upon a plan of necromancy. The priest
one evening prepared to satisfy me, and desired me to look out for
a companion or two. I invited one Vincenzio Romoli, who was
my intimate acquaintance; he brought with him a native of Pis-
toia, who cultivated the black art himself. We repaired to the
Collosseo, and the priest, according to the custom of necromancers,
began to draw circles upon the ground, with the most impressive
ceremonies imaginable; he likewise brought hither asafœtida, several precious perfumes, and fire, with some compositions also, which
diffused noisome odors. As soon as he was in readiness, he made
an opening to the circle, and having taken us by the hand, ordered
the other necromancer, his partner, to throw the perfumes into the
fire at a proper time, intrusting the care of the fire and perfumes
to the rest, and thus he began his incantations. This ceremony
lasted above an hour and a half, when there appeared several legions
of devils, insomuch that the amphitheatre was quite filled with
them. I was busy about the perfumes, when the priest, perceiving
there was a considerable number of infernal spirits, turned to me
and said, "Benvenuto, ask them something." I answered, "Let
them bring me into the company of my Sicilian mistress, Angelica."
That night he obtained no answer of any sort; but I had received
great satisfaction in having my curiosity so far indulged. The
necromancer told me it was requisite we should go a second time,
assuring me that I should be satisfied in whatever I asked; but
that I must bring with me a pure immaculate boy.

"'I took with me a youth who was in my service, of about twelve
years of age, together with the same Vincenzio Romoli, who had
been my companion the first time, and one Agnolino Gaddi, an intimate acquaintance, whom I likewise prevailed on to assist at the
ceremony. When we came to the place appointed, the priest having made his preparations as before, with the same and even more
striking ceremonies, placed us within the circle, which he had likewise drawn with a more wonderful art, and in a more solemn manner than at our former meeting. Thus, having committed the care
of the perfumes and the fire to my friend Vincenzio, who was
assisted by Agnolino Gaddi, he put into my hand a pintaculo or
magical chart, and bid me turn it towards the places that he should

direct me; and under the pintaculo I held the boy. The necromancer, having begun to make his tremendous invocations, called by their names a multitude of demons who were the leaders of the several legions, and questioned them, by the power of the eternal uncreated God who lives forever, in the Hebrew language, as likewise in Latin and Greek; insomuch that the amphitheatre was almost in an instant filled with demons more numerous than at the former conjuration. Vincenzio Romoli was busied in making a fire, with the assistance of Agnolino, and burning a great quantity of precious perfumes. I, by the directions of the necromancer, again desired to be in the company of my Angelica. The former thereupon turning to me, said : "Know, they have declared, that in the space of a month you shall be in her company."

"'He then requested me to stand resolutely by him, because the legions were now above a thousand more in number than he had designed; and besides, these were the most dangerous; so that, after they had answered my question, it behooved him to be civil to them and dismiss them quietly. At the same time the boy under the pintaculo was in a terrible fright, saying that there were in that place a million of fierce men, who threatened to destroy us; and that, moreover, four armed giants of enormous stature were endeavoring to break into the circle. During this time, whilst the necromancer, trembling with fear, endeavored by mild and gentle methods to dismiss them in the best way he could, Vincenzio Romoli, who quivered like an aspen leaf, took care of the perfumes. Though I was as much terrified as any of them, I did my utmost to conceal the terror I felt, so that I greatly contributed to inspire the rest with resolution; but the truth is, I gave myself over for a dead man, seeing the horrid fright the necromancer was in. The boy placed his head between his knees and said, "In this posture will I die, for we shall all surely perish." I told him that all these demons were under us, and what he saw was smoke and shadow; so bid him hold up his head and take courage. No sooner did he look up than he cried out, "The whole amphitheatre is burning, and the fire is just falling upon us." So covering his face with his hands, he exclaimed, "that destruction was inevitable, and desired to see no more." The necromancer entreated me to have a good heart, and take care to burn proper perfumes; upon which I turned to Romoli, and bid him burn all the most precious perfumes he had.

At the same time I cast my eye upon Agnolino Gaddi, who was terrified to such a degree that he could scarce distinguish objects, and seemed to be half dead. Seeing him in this condition I said, " Agnolino, upon these occasions a man should not yield to fear, but should stir about and give his assistance, so come directly and put on some more of these." The effects of poor Agnolino's fear were overpowering. The boy hearing a crepitation, ventured once more to raise his head, when, seeing me laugh, he began to take courage, and said "that the devils were flying away with a vengeance."

" ' In this condition we stayed till the bell rung for morning prayers. The boy again told us that there remained but few devils, and these were at a great distance. When the magician had performed the rest of his ceremonies, he stripped off his gown and took up a wallet full of books which he had brought with him.

" ' We all went out of the circle together, keeping as close to each other as we possibly could, especially the boy, who had placed himself in the middle, holding the necromancer by the coat, and me by the cloak. As we were going to our houses in the quarter of Banchi, the boy told us that two of the demons whom we had seen at the amphitheatre went on before us leaping and skipping, sometimes running upon the roofs of the houses, and sometimes upon the ground. The priest declared, that though he had often entered magic circles, nothing so extraordinary had ever happened to them.

" ' Whilst we were engaged in this conversation, we arrived at our respective houses, and all that night dreamed of nothing but devils.'

" Although Cellini declares that he was trembling with fear, yet it is quite evident that he was not entirely ignorant of the machinery which was at work, for in order to encourage the boy, who was almost dead with fear, he assured them that the devils were under their power, and that ' what he saw was smoke and shadow.'

" Mr. Roscoe, from whose life of Cellini the preceding description is taken, draws a similar conclusion from the consolatory words addressed to the boy, and states that they ' confirm him in the belief that the whole of these appearances, like a phantasmagoria, were merely the effects of a magic lantern produced on volumes of smoke from various kinds of burning wood.' If we suppose that the necromancer either had a regular magic lantern, or that he had fitted up his concave mirror in a box containing the figures of his devils, and that this box with its lights was carried home with the party, we can easily account for the declaration of the boy, 'that as they were going home to their houses in the quarter of Banchi, two of the demons whom we had seen at the amphitheatre went on before us leaping and skipping, and sometimes running upon the roofs of the houses, and sometimes upon the ground.' "

We could hardly, in this enlightened age, attain to the brilliant success of frightening a "pure immaculate boy" out of his senses with "smoke and shadow," even were it a laudable undertaking. The delirium tremens, in a somewhat similar way, will doubtless continue to be hard on older and wayward boys who take to their cups, but be it ours to please and instruct, and that, in a more excellent way. A jet of steam could be conveniently arranged for the "ghost" experiment, but for the most part, a wet screen is better than smoke, and effects, not only startling, but truly beautiful, can be produced in the way described.

CHAPTER IV.

PICTURE SLIDES.

A LARGE number of movable slides, and some others of value, are still painted entirely by hand, but the great part of simple slides now in market are produced by photography.

There are two classes of photographic transparencies for the lantern, viz.: instantaneous and other views direct from nature, and reproductions of ancient and modern engravings, or paintings. A great part, especially of the latter, are beautifully colored by skilful artists, and mounted in a round form in wooden frames.

Some idea of the value of photography, associated with the magic lantern, as an educational instrument, may be gathered from the fact that as the camera has now penetrated to almost every habitable part of the globe, the physical peculiarities of every country, together with lifelike portraits of their inhabitants, and the form and arrangement of their dwellings, may be obtained in miniature, and reproduced as large as life.

Photographs of the sun and moon in various phases, and partially and totally eclipsed, also the fixed stars and nebulæ, have been obtained and employed for lecture illustrations. Enlarged photographs of microscopic objects have also been obtained, and these again still further enlarged to 8 or 10 feet in diameter, so that, in fact, a diatom no larger than a grain of sand may be shown of such a size in the lecture-room that a large audience may together examine its details with perfect comfort. The productions of the most celebrated painters and sculptors may be shown with equal facility, as well as maps, hymns, music, &c., so that an entire school may learn or sing together.

THE STANDARD SIZE FOR LANTERN SLIDES.

The ordinary wooden frame for the lantern picture is 7 inches long, 4 inches wide, and ⅜ of an inch thick, with a circular opening of 3¼ inches to admit the picture-glass and its protecting glass cover, and 3 inches in the clear. Pictures 3¼ inches square are also mounted in frames of the same size, leaving 3 inches square in the clear. Pictures 3¼ inches square, with their protecting glass covers, are also bound with narrow binding, and may be slid along into place in the grooves of a stationary frame, so as to show 3 inches square.

The new Woodbury slides are exactly quarter plate size (3¼ x 4¼), which gives room at the sides for naming and for handling, and which in turn gives assurance of their being inserted in proper position.

Fortunately each half of a stereoscopic view is 3 inches square, so that lantern slides, of standard size, can be printed by contact from stereoscopic negatives. Although the demand for lantern slides has never warranted extensive travel for desirable negatives, yet the stereoscope has sent photographers "viewing" high and low, and everywhere; on the Alps, in the Yosemite, in the valley of the Nile, on open Polar Seas, and often (as intimated by one of their own number) into distressingly *narrow straits.*

Glass transparencies made for the stereoscope, when cut in two, with clear glass covers instead of ground-glass, are extensively used for lantern slides. Many of these, especially of the imported views, are very fine, and leave nothing to be desired when used in the Sciop-ticon. But as a heavy deposit of silver is not particularly objectionable in the stereoscope, many of these pictures can only be satisfactorily shown upon the screen, with an intense chemical light, if with that even. When the demand for these fine views for the lantern is suffi-cient to turn the attention of photographers to their production, we may look for more good pictures, and we hope at a cheaper rate.

THE SEALED PICTURE.

A drop of Canada balsam between two disks of plate-glass, on one of which is a beautifully colored photo-graph, is skilfully managed so as to allow the plates to come almost in contact, with a film of the balsam filling all the space between; this makes the picture beautifully

clear, and protects it from dust, and especially from any
fluid that might accidentally reach the edges of the glass.
All the circular pictures, except the movables, at a price
of over two dollars, are scaled.

STATUARY.

This class of pictures should be photographed directly
from the statuary, or bas-relief, by a skilful artist, who

Fig. 18.

understands lighting and how to secure the proper
degree and gradations of shading. It appears to the

best advantage when the rest of the slide is made
opaque, so that we may seem to lose sight of the screen,
and see the figure standing out in open space.

The circle, including Thorwaldsen's Night (Fig. 18),
shows the size of the round glass, and also the appearance of one of the most popular slides of this class.
A female figure is seen floating down to earth; around
her forehead is a wreath of poppy, indicating sleep; in
her arms are two sleeping children (Sleep and Death);
and in their company is the symbolic owl.

THE SLIP SLIDE.

Fig. 19 represents a class of movable slides most in
use for amusement; being cheap, easily operated, and
in shape to pack with ordinary slides.

In the slide represented, a peacock without a tail is
painted on the immovable glass, and two tails are painted

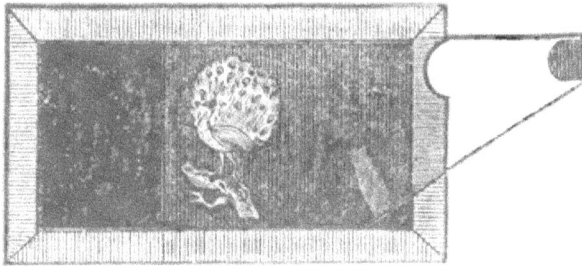

Fig. 19.

on the slip. Both glasses are blackened except where
the picture is to show; when the slip is pushed in, the
bird appears with a drooping tail; when the slip is drawn
out, then you will see him spread.

In some of these there is a slip each side of the fixed
glass. In the popular slide called the rat-eater, a man

reclining on a bed is painted on the fixed glass, a lower jaw on one slip, which works up and down, and a rat on the other, which, as the slip is drawn, has the appearance of running down the man's throat. With the Sciopticon, the operator has hold of a slip with each hand, so he can jerk the rat back with a sudden movement of the forefinger, when he is all ready to make his appearance again as a new individual. In politics he might be called a "repeater."

THE LEVER SLIDE.

Fig. 20 represents another popular, but a more expensive, mechanical effect. The horse having approached the water with his head up, the lever to the right is raised, and the horse is *"made to drink"* (the old adage to the contrary notwithstanding). The head and neck

Fig. 20.

being painted on the glass moved by the lever, works up and down as on a pivot at the shoulders.

REVOLVING FIGURES.

A movable disk corresponding to the one moved by the lever in Fig. 20, may be revolved by means of a rack and pinion; of this class is a variety of chromatropes, mill-wheels, the movable astronomical diagrams, &c. Suppose a mill and the surrounding landscape painted on the fixed disk, and the sails on the disk revolved by rack and pinion, then on turning the handle, the mill appears in operation.

THE CHROMATROPE.

Fig. 21 represents the pulley form of the chromatrope, but can give no idea of the dazzling brilliancy of the effects it produces on the screen. There is nothing it resembles so much as the kaleidoscope, with the addition of constant motion and rapid change. It consists of two disks of glass, painted with an almost endless variety of geometrical and other designs in brilliant colors. By turning the handle shown in the figure, the multiplying band causes the rapid revolution of one disk over the other, producing two apparent motions; and with good designs the result " beggars all description."

Fig. 21.

In another form the motion is accomplished by means of a double rack and pinion, instead of a band. As there are two disks, revolving one over the other, each is necessarily furnished with a toothed rack, and the one pinion works both.

THE EIDOTROPE.

This valuable accessory to the magic lantern consists of two perforated metal disks, which, by an arrangement precisely similar to the chromatrope, are made to revolve one over the other. The effects are so beautiful as to excite surprise that they should be obtained by a mechanical contrivance of such extreme simplicity. By slow revolution, hexagonal, octagonal, and other geometrical figures are obtained, with delicate gradations of shadow; while a more accelerated motion produces the effect of stellate flashings, or scintillations of light. Color may be imparted by the use of tinted films of gelatine. Larger disks can be pivoted to a frame above the condenser so as to give an upward and outward movement to the scintillations as from a lower fountain. Stiff paper disks turned contrary ways by the hands at their edges will answer for practical experiments.

MOVING WATERS.

Under this title two forms of slides are sold; by means of which, in a single or double lantern, very pleasing effects may be produced. In the simplest form a moonlight scene is painted on a fixed disk, and the "rippling waters" on a piece of glass attached by one corner only to the framework of the slide, which being moved up and down causes the appearance of a ripple on the water.

Another more expensive, but more truthful effect, is produced by a slide having two movable and one fixed disk of glass, and known as the "moving water with eccentric motion." In this slide, not only is the ripple produced, but the heaving of a boat upon the waves, the "rolling" of the sea, and the "hovering" motion of birds is imitated with surprising closeness to nature.

5

LONG SLIDES.

A very large class of paintings, on strips of glass 12 or 14 inches long, have come down to us from a former generation. Many of them hardly deserve attention, but some Nursery Tales, Natural History, &c., are fair, and the Astronomical set, in particular, is excellent.

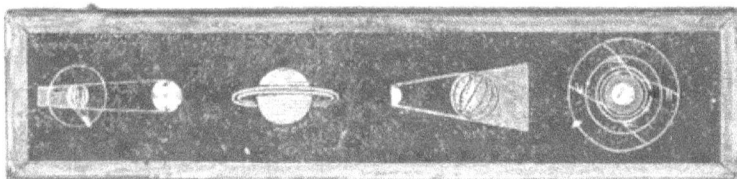

Fig. 22.

This set of ten astronomical slides, with forty-one illustrations, together with a set of astronomical diagrams with rack-work motion, makes a very complete outfit for a series of astronomical lectures.

DIORAMIC PAINTINGS WITH MOVING FIGURES.

In the middle of a glass strip (shaped and framed as in Fig. 22), a scene is painted, the rest being made opaque. Another glass strip, of similar size and shape, on which is painted along its whole length whatever is befitting, as figures, boats, &c., is made to pass in front in grooves, so as to represent a long procession ; of this class, the children of Israel passing through the Red Sea is an example ; or, the enterprising smugglers secreting contraband goods in the smugglers' cave.

THE GHOST.

A description of the modern "ghost" may be looked for here, but it is not strictly a magic lantern production.

A large plate of glass leans forward on the front of the stage, but its edges are so hidden by the curtains that its presence is not suspected. A "woman in white" stands down in front, concealed from the spectators by the usual board near the orchestra, and is highly illuminated by the light from a magic lantern. The spectators, in the darkness and distance, see the actors upon the stage *through* the glass, and also the *ghost* reflected *from* the glass so as to appear on the stage with the rest. The actors do not see the spectre, but they put on the appearance of fright for the *benefit* of the spectators.

The apparition vanishes as the light is withdrawn from "the woman in white." The lantern is used because it illuminates an object without diffusing light in other directions.

On this principle we may see people in a room through a window, with the reflected images of parties outside standing among them. It seems not a little surprising to see one person cutting through the space occupied by another.

THE TANK.

An excellent and cheap tank (similar to the one shown in Fig. 26), but with permanent clamps without screws, is now shaped so as to slide into the Sciopticon stage without drawing forward the extension front. As the space at the top is unobstructed, all sorts of experiments with it are easily managed. Living creatures encaged in it, in air or water, figure upon the screen in huge proportions, and with wonderful activity. Some fish and parts of many insects are so transparent as to show internal structure. Even opaque objects, when their outlines are sharply focused, appear in relief unlike a mere shadow. With almost every object thus shown, except fish, up and down is a matter of indifference; so that inversion is no disadvantage. With this tank for the exhibition of living objects, chemical reactions, &c., a large assortment of slides is less of a necessity.

PRECAUTIONS ABOUT SLIDES.

The lantern exhibition has to be conducted in so obscure a light that the operator has to depend more on the sense of feeling than sight; it is therefore important that the slides should be in good condition and properly arranged beforehand, and that their titles and descriptions should be well fixed in memory.

A convenient box for carrying the slides, for arranging them in, and for showing them from, is constructed as follows: Two boxes of any desirable length, 7¼ inches wide and 4¼ inches deep, are hinged together, so that each serves as a cover to the other. This double box will hold the ordinary wooden mounted slides without waste of room, and when open will show their labelled edges in proper position and order.

In social gatherings, the exhibitor is often urged to bring out certain favorite pictures on call, which, in the hurry and darkness, is apt to disarrange the slides, so as to perplex the operator, and mar the beauty of the entertainment.

With careful management the box may close on properly arranged slides, at the close of the exhibition.

CHAPTER V.

𝔓𝔥𝔬𝔱𝔬𝔤𝔯𝔞𝔭𝔥𝔶.

GLASS POSITIVES FOR THE MAGIC LANTERN.

BY JOHN C. BROWNE.

FEW entertainments for the amusement of children, as well as persons of mature years, give more real pleasure than exhibitions of the magic lantern. It is a never-ending source of pleasure, and doubly valuable to the disciple of photography, who by the aid of a few chemicals and very simple apparatus, can prepare interesting slides of local interest that will delight the home circle, and fully repay the small expenditure of time required for their manufacture. Every photographer has among his negatives many subjects, both portrait and landscape, that when printed upon glass will prove effective pictures for exhibition.

The object of this paper is to give in as few words as possible, plain directions for making positives on glass, suitable for the magic lantern.

Either the wet or dry process can be used. The former is more applicable in cases where it is necessary to reduce a negative to the proper sized positive required for the lantern. The dry method is used to advantage when the negative is of small size, and can be printed in contact. As all photographers are familiar with wet manipulations, we will consider that process first.

The only apparatus actually required, is an ordinary camera and lens, placed upon a board six feet long, in front of which a negative is fastened. This negative is simply copied upon a sensitive collodion plate, that is exposed in the camera, carried into the dark-room, and developed, fixed, and toned.

If many positives are to be made from negatives of different sizes, it will be found convenient to arrange two cameras front to front (as will be seen in Fig. 23),

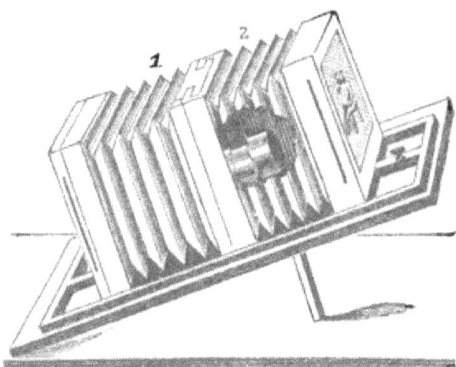

Fig. 23.

one camera having a lens in position with ground-glass No. 1; the other having the lens and ground-glass removed, and the negative that is to be copied placed in the position of ground-glass No. 3. The cameras, for convenience, can be closely joined together by screwing strips of wood upon each, which prevents any change of position when focused. By turning the rack-work upon each box, the picture upon the ground-glass can be made of any size.

This plan of using two boxes will be found to give perfect satisfaction; the negative and sensitive plates are always on the same plane, and the adjustment for difference of size made in a moment. It is only for convenience of manipulation, that the writer advocates this arrangement, for excellent positives can be obtained by using one camera, and placing the negative to be

copied in proper position in front of the lens, with nothing to shut off the diffused light between the lens and
negative.

In selecting a position for operations, a north light
will be found the best suited for the purpose. The
upper window sash should be lowered, and the board
upon which the cameras are arranged rested one end
upon a table, the other upon the lowered sash, so that
the negative will have the sky for a background. This
is easily determined by focusing upon the ground-glass.
It is advisable not to allow the direct rays of the sun to
illuminate the negative. Should a north light not be
obtainable, cover a frame with white tissue-paper, and
let that be the background for the negative. The tissue-paper can be illuminated with the sun's rays, or by
any artificial light.

It may seem to the reading photographer, unneces-
sary to burden this article with a complete chemical
formula for making glass positives, but as it is prepared
expressly for the uninitiated, it would be unintelligible
without a formula.

To make 8-oz. Sensitive Collodion.—Alcohol 5 oz., ether
3 oz., iodide of ammonium 44 grs., bromide of magnesium 20 grs., cotton (Parys') 35 grs. Before using, filter
several times through cotton soaked in alcohol. It is a
good plan to keep a supply of plain, unexcited collodion
on hand, as a stock-bottle; also, a bottle of exciting solution, made in the proportion of iodide of ammonium 5
grains, bromide of ammonium 2½ grains, to the drachm
of alcohol. By adding 1 drachm of the solution to 1
ounce of plain collodion, it will be excited to the proper
condition.

Nitrate Solution.—Water 1 ounce, nitrate of silver 40
grains; make slightly acid with nitric acid, C. P. Iodize

the solution by allowing a plate coated with excited collodion to remain in it over night.   Filter.

Developing Solution.—Make a **saturated solution of**
ammonia, **sulphate of iron (in water); filter.   To every**
ounce of this solution add glacial acetic acid, 1 drachm.
This **can be used as a** stock **solution, and will** keep an
indefinite length of time in good condition.   Crystals
will form in the stock-bottle, after standing some **hours,**
but that is of no consequence, as the strength of the
solution is correct.

In developing a plate, use 5 drachms of water **to 2**
drachms of ammonio-sulphate of iron from the **stock-**
bottle.   During hot weather use ice-water to retard the
action **of the** developer.

Fixing Solution.—Cyanide of **potassium or hyposul-**
phite of **soda; either will answer, but the action of**
cyanide appears to make a somewhat brighter **picture.**

These solutions being carefully prepared, the picture
accurately focused, the negative (collodion side towards
the lens) covered **with a** dark cloth, prepare the plate
in the dark-room in the usual manner, place it in the
dark-holder, in the position of the ground-glass, draw
the slide (the lens is always uncovered), remove the
cloth from the negative for a few seconds.   The exposure will then be made.   **Cover the negative,** shut the
slide, and remove to the dark-room for development.
The picture should appear **slowly; not flash out upon**
the first application of the iron **solution.   Over-expos-**
ure, as well as over-development, are both fatal **to trans-**
parencies.   No trace of fog should be visible.   From
five to fifteen seconds will be found sufficient, on **a bright**
day, with a negative of ordinary strength, and the
chemicals in **good order.**

The Negative—Must **be *sharp*,** of good printing density, and as free as possible from all defects. As the magic lantern slide is generally 3¼ x 3¼, it is not desirable to employ a very large negative. 6½ x 8½ will answer the purpose better than a larger size. But negatives upon smaller glass will be found to give even finer results. On the other hand, it is very bad policy to attempt to enlarge a positive to double or treble the size of the original negative. The negative should not be smaller than the positive.

The Lens.—Any good portrait combination, of six to eight inches focus, quarter-inch stop, will work to advantage. Lenses of very short focus and very small opening, are not recommended.

The Development—Should be conducted with great care and judgment, as it is the most important part of the whole process. Rather underexpose and underdevelop, and as soon as the detail is visible, flood the plate with water, and check further action. Avoid an excess of light during development, and dread the appearance of the slightest fogging as the worst enemy to be encountered.

Fixing Solution.—Cyanide of potassium, after which wash well in running water.

Toning.—It is frequently of benefit to the positive that it should be toned, and at the same time slightly strengthened, to give contrast to the picture when projected upon the screen by a powerful light. Many chemical solutions may be used to accomplish this purpose. A weak solution of gold gives good results; also, a dilute solution of bichloride of palladium can be recommended. In either case the solution is flowed over the plate, after fixing. The positive is then dried and varnished.

The Finished Picture—Should be free from the slightest appearance of fog; the high-lights, the sky in land-

scapes (except when clouds are present), perfectly clear
glass. The particular tone requisite to suit the positive,
is a matter of taste. A warm sepia will be found suitable for most transparencies; but each operator must
exercise his own peculiar feeling in this matter.

In making positives to be exhibited by the magic lantern, it is well to consider the variety of light to be used
in projecting the picture upon the screen. Where powerful illumination, such as the oxy-hydrogen or magnesium lights are used, positives may be made slightly
stronger, showing more contrast than where a weaker
form of illumination is employed.

The slides should be protected from scratches and
dust, by a piece of clear glass of the same size, neatly
pasted on the edges with muslin.

Positives on glass can also be made by the wet process, from negatives of the proper size, by pasting a thin
strip of cardboard upon two edges of the negative (collodion side). The sensitive plate is prepared as usual,
and is placed, while in the dark-room, in close contact
with the negative, separated only by the cardboard. It
is then exposed behind the negative, to diffused sunlight
or artificial light, for a few seconds, returned to the dark-
room, and developed. This plan admits of no change
in the size of the negative. Mr. L. J. Marcy's apparatus for printing wet plates by lamp-light, has given satisfaction to many who have not an opportunity of making experiments by daylight.

The proper size for glass pictures to be used in lanterns of convenient proportions, is a debatable subject.
Glasses of 3¼ x 3¼ being generally used, but advantages
are claimed for a slide 3¼ x 4¼, that have some weight.
In placing this slide in the lantern, the additional length
of the glass allows the corners to be held by the thumb

and forefinger, without being visible upon the screen, as is sometimes the case with the square slide. Then again, it is easier to place in its proper position (right side up), having only one chance of error instead of three.

A DRY PLATE PROCESS FOR LANTERN SLIDES.

TANNO-GALLIC PRESERVATIVE.*

In considering the dry process, it is but proper to say that a large number of different formulæ have been published; in fact, scarcely half a dozen photographers think alike on this subject. It is, therefore, impossible to give a formula that will give universal satisfaction. In preparing this paper for publication, it must be distinctly understood that nothing new in the way of preservative or development is claimed; it is simply one of the many methods for preparing dry plates that has given reliable results.

The dry-plate photographer must be prepared for many and great failures, and be possessed of the greatest amount of patience and nicety of manipulation, for otherwise time is wasted, and the best process voted a failure. Commence with reliable chemicals, and follow up the process with a lavish expenditure of water when washing is mentioned, not only on the collodion plate, but thoroughly rinse the various glasses and dishes, and particularly the *fingers*, between each operation. Use as little light as possible when making or developing dry plates, and be careful that the light is yellow.

Probably more dry plates are ruined, and the particular process used condemned as worthless, by the use

---

* I am greatly indebted to my friend, Mr. E. Wallace, Jr., for his kindness in furnishing me with the formula, and showing, by his own experiments, the valuable results to be obtained from this process.   J. C. BROWNE.

of white light than from any other cause. A square-sided lantern, having the white glass removed, and yellow substituted, will be found very convenient; either gas, a candle, or kerosene can be used for illumination.

To prevent the collodion film slipping from the plate during the process, it is absolutely necessary that the glass plate should be albumenized. Wash the glass (having previously roughened the edges), drain, and while wet flow over it the following solution:

Albumen (the white of an egg), ... 1 egg.
Water, ... 1 pint.
Concentrated Ammonia, ... 10 drops.

Put the albumen in a clean bottle, then add the water. Shake a little, and add the ammonia; filter through a sponge; dry in a rack.

COLLODION.

Any reliable collodion will answer; it is best to have it quite thick. No backing is necessary.

NEGATIVE BATH.

Nitrate of Silver, ... 45 grains.
Water, ... 1 ounce.

Made slightly acid with nitric acid, C.P. Dip the collodionized plate in the bath, and when properly excited, remove the plate, and dip in a bath of pure water; then wash under a tap with running water. While wet apply the

PRESERVATIVE SOLUTION.

Tannin, ... 10 grains.
Gum Arabic, ... 6   "
Sugar, ... 4   "
Water, ... 1 ounce.

Filter, and add one drachm per ounce of

Gallic Acid, ... 24 grains.
Alcohol, ... 1 ounce.

The preservative must be fresh. Three ounces of this mixture will prepare half a dozen 6½ x 8½ plates.

If the preservative is poured over the plate, apply twice, working it well into the film; throw the first *dose* away, and use the second flowing for the first application to the next plate.

The plates must be carefully dried, either by natural or by artificial heat; a hot-water bottle will be found useful for that purpose should artificial heat be thought best.

THE EXPOSURE

Will depend upon the strength of the negative, and the nature of the light; a few seconds will generally be enough. *Close contact* is absolutely required to produce sharp positives. An ordinary printing-frame can be used.

TO DEVELOP

In a dark-room, remove the dry plate from the frame, place it in a dish, and flow over it

Alcohol, ... } equal parts.
Water, ... }

Then wash in running water.

DEVELOPING SOLUTIONS.

Pyrogallic Acid, ... 2 grains.
Water, ... 1 ounce.

Made from a stock-bottle of

Alcohol, ... 1 ounce.
Pyrogallic Acid, ... 96 grains.

Five minims of this solutions contains one grain of pyro.

Carbonate of Ammonia, ... 64 grains.
Water, ... 1 ounce.
Bromide of Potassium, ... 4 grains.
Water, ... 1 ounce.

Mix together.

After the plate is well washed, flow over it a solution of

Pyrogallic Acid, ... 2 grains.
Water, ... 1 ounce.

Then pour back again into the measure. Should the image be developed by this solution, proceed very cautiously, and add a few drops of the alkaline solution of carbonate of ammonia and bromide of potassium. If the picture comes out slowly, add more of the alkaline solution up to thirty drops, if necessary, and also a sufficient amount of stronger pyro to bring out all the detail. When the image is out, wash with water, and intensify with

Pyrogallic Acid, ... 2 grains.
Water, ... 1 ounce.

To which is added ten drops of citric acid and nitrate of silver solution.

Citric Acid, ... 30 grains.
Nitrate of Silver, ... 20   "
Water, ... 1 ounce.

This is a stock-bottle. Mix in separate glasses; add together and filter; wash.

Hyposulphite of soda.

The same remarks applied to wet positives will answer for toning dry plates.

THE COLLODIO-CHLORIDE PROCESS.

(From Humphrey's Journal.)

The following formula is not only used for opal pictures, but to some extent for transparencies also.

1. Take the whites of two eggs and two ounces of water, beat well to a froth, and let it settle for two hours and pour off the clear solution.

2. Coat your white plate with this solution (as you would with collodion), and set away to dry. When dry take in your dark-room and coat the plate with the "opal solution," which is made thus:

Plain collodion 8 oz. (thinner than you would use for iodizing), then dissolve in as little water as possible 60 grains nitrate of silver, and add this to the collodion and shake well. Then dissolve 16 grains of strontium in as little water as possible, and add this to the collodion and shake well. Then dissolve 10 grains citric acid in as little water as possible, and add to the collodion. Shake well, and you have the opal solution.

When dry, put your negative in the printing-frame, lay the opal-prepared plate on the negative, and print from 10 to 15 minutes in the sun, and print much darker than you would a photograph.

Tone and fix as you would a photograph, only you need not wash before toning—and wash but little before fixing. The "opals" tone in one-tenth the time of a photograph.

Keep the opal preparation in a dark-room. Have your toning bath a little alkaline, and not as strong as for toning photographs.

MARCY'S PHOTOGRAPHIC PRINTING APPARATUS FOR PRINTING WET PLATES BY LAMPLIGHT.

This apparatus is intended to simplify the process of printing lantern transparencies. Its rationale will be seen at a glance.

Sharp photographic printing without a camera, can be effected, either *by having the negative in actual contact with the sensitive plate*, however widespread the light, or else *by having an intense light proceeding from a single point*, though the plates may be wide apart. In the latter case the point of light should be distant compared with the space between the plates, to avoid enlargement. A sharpness above criticism is produced by this printing apparatus, not by an absolute compliance with either condition, but by an approximate observance of both.

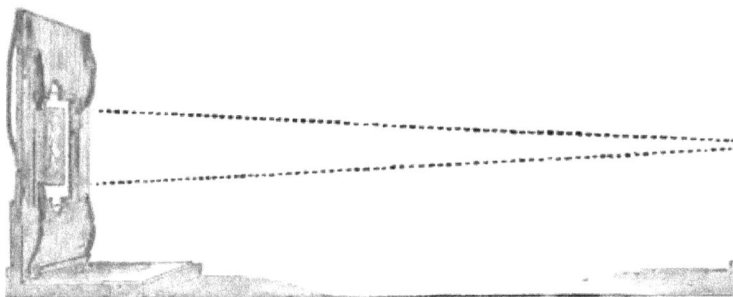

Fig. 24.

It consists of an upright frame in which the sensitive plate is held slightly separate from the negative, and a coal oil lamp, from which the light of a wide flat flame is emitted through a narrow horizontal slit—small and at considerable distance from the frame to produce a sharp print, and in range with the long diameter of the flame to get intensity from a single point. A narrow strip of glass sets into this slit as between two lips. The thickness of the flame gives the horizontal diameter of the point of light. Only the front of the lamp is shown at the right of Fig. 24, but it can be seen how the light from the whole width of the flame reaches the printing-frame through the narrow aperture.

At the left, we see how the negative is held over the opening in the frame by four springs; the long spring on the opposite side holds the sensitive plate in the frame.

The operator, standing on the opposite side, with the upper corners of a quarter plate, just from the bath, between his thumb and finger, and shading off direct rays with his left hand, places it in its silver bearings; this brings the two films almost in contact.

The lamp and frame stand from 16 to 26 inches apart, or so far as to require about two minutes for the printing, or the time it takes for a round of the other manipulations and changes; so a picture is finished and dropped into grooves in a trough of water as often as one has had time to print.

All that is said in the previous article on the wet-plate process, in regard to development, &c., applies here. Any drops of silver bath that may have come in contact with the negative must be washed off before it is put away.

Like dry-plate printing, the negative must be of the exact size required for the lantern slide.

Some of the advantages of this method are:

1. It can be practiced evenings or in cloudy weather.
2. The light is inexpensive.
3. Plate glass is not a necessity.

4. The apparatus may stand within reach of the operator.

5. It requires no previous preparation more than having clean glass, and chemicals in good working order.

6. The albumen coating is not required to make the film adhere.

7. The amount of exposure can be definitely gauged.

8. The illumination is confined to a narrow cone, so as not to fog the picture by diffused light.

9. The exposure is so immediate and uniform as to escape many accidents.

10. It is so easily done, that many causes of failure involved in a long process are not encountered.

11. The negative is not marred by use as in contact-printing.

12. This apparatus complete costs but seven dollars.

Thus we have in it advantages by the dozen.

THE SCIOPTICON PROCESS.

By placing the Sciopticon near a wall, in a dark room, and drawing forward its extension front, an image of a negative may be projected into a three-inch circle. First focus sharply on a paper-covered glass, and then expose a wet plate in the same place a minute, more or less, developing and fixing as usual, and we have a glass positive photographed by the Sciopticon for the Sciopticon. The objective is always used with full opening, because all the light is needed, and because it will not give an evenly illuminated disk with a small stop; so we cannot secure perfect sharpness to the very edges. It answers well, however, for central figures, and the photographer can easily produce unexceptionable positives from his

portrait negatives. There seems no reason why enlarge-
ments made in this way, for *ornamental transparencies*,
to be hung in the window, or set in a frame, should not
become a profitable branch of photography.

The toning of glass positives, to be used for ornamental purposes, involves some thought as to the particular color, or shade of color, that will suit the picture
best; and it is impossible to give one process that will
suit all tastes alike; some having a preference for black
tones, others for blue-black, brown, or the various shades
of gray. A detailed description of the manner of producing these various tones would require too much space,
and is so simple that no one can go astray.

The principal chemicals required are: Chloride of
gold, bichloride of platinum, bichloride of palladium,
sulphide of potassium, and permanganate of potassium;
in all cases use singly and very dilute. I am disposed
to consider chloride of palladium as the most reliable
chemical that has come under my notice. Its action is
perfectly manageable, easy to prepare, will not stain,
and gives uniformly good results. The toning solution
that I use is made as follows : Add six drops from the
stock-bottle of chloride of palladium to each ounce of
water; this solution should be of a delicate straw color.
No other manipulation is required. After the plate has
been developed and fixed, wash as usual, then apply the
toning solution by flowing it over the plate similarly to
the developer. Its action will be quick, giving a black
tone to the positive. Wash well, dry, and varnish if
desirable.

These transparencies are covered and bound with an
opal or ground-glass, or they may be flowed with a varnish containing a little fine zinc paint, ground in varnish.

It may be mentioned in this connection, that artists

find the Sciopticon very useful in *sketching their pictures*. Having first obtained a glass positive or negative of the subject to be painted, it can be thrown upon the canvas of the size desired, and expeditiously and accurately traced. It saves valuable time to the good artist, and it prevents the poor artist from producing distortions.

WOODBURY PHOTO-RELIEF EXCELSIOR LANTERN SLIDES.

By John C. Browne.

While it is a comparatively easy matter to produce fine positives by either the wet or dry process of photography, yet the results are liable to vary somewhat even in the hands of the most careful manipulator. The Woodbury photo-relief process, as now worked in Philadelphia, has the merit of distancing all competition in the uniform excellence of its lantern slides. It would be a pleasure to give in detail a description of this wonderful process, did space permit, commencing with the sensitive gelatine tissue, resembling in appearance a piece of patent leather, and following it in its exposure to light under a negative, the light's action rendering insoluble those parts reached through the negative; its subsequent immersion in hot water dissolves out those parts not rendered insoluble, producing a relief as thin as writing paper, which when dry is pressed into a piece of soft metal by a hydraulic press of fabulous power, forcing this delicate substance into the smooth metal, and leaving upon its surface a counterpart or mould of all its finest lines and half tones. Strange to say this flimsy gelatine relief is not crushed to atoms by this treatment. It is not damaged in the least, but ready to make its mark again as often as it is necessary.

This leaden mould is the type that prints the picture, a solution of gelatine and India-ink being poured over it before the glass is placed in position.

A slight pressure is given in a press of peculiar construction, squeezing out the surplus ink; a few minutes is allowed the ink to set, when the glass, being removed, brings with it the delicate gelatine picture, which is well named " Excelsior."

CHAPTER VI.

COLORING SLIDES.

WRITTEN FOR THE SCIOPTICON MANUAL.

THE magic lantern has caused much astonishment and delight from its origin to the present time. The pictures or slides for it were formerly drawn or painted on glass, and when magnified by the lantern lens, even the most minute lines looked coarse, and every imperfection was brought out. Much time and care, therefore, were requisite to make fine pictures, so that they were comparatively rare and expensive, while the coarser ones abounded; thus the lantern came to be regarded as a toy, fit only for the amusement of children. An instrument, however, so well calculated to aid in the advancement of science and education, on account of the size of the diagram that may be represented on the disk, and the fact that the attention of an audience is better secured when the only object visible is the dia-

gram under explanation, was not to be thrown aside as
a toy.

Photography, by its wonderful sun paintings on glass,
reproduces the works of the old masters, furnishes views
of every land and clime, of customs, manners, works of
art, and pictures, or diagrams, to illustrate every science,
the beauty of which, when colored and thrown upon the
screen, however great the magnifying power used, is not
diminished, as was the case with the paintings formerly
used. With beautiful and desirable pictures, and with
improved lights and instruments, the lantern now takes
a front place in Sunday-school work, in the school, the
lecture-room, and the home, and is gladly welcomed
wherever visible illustrations are used, or beautiful pictures prized. While the stereoscope presents the life-
like photographs to the individual observer, the lantern
enlarges the same views, so that many may see and enjoy
at the same time the same beautiful scenes together,
making it well suited to the social gathering and entertainment of friends. Families may have slides prepared
containing pictures of family residences, of members of
the family, of favorite dogs, horses, &c., thus increasing
the pleasures of home, and social intercourse.

The coloring or painting of slides for the magic lantern has been confined to comparatively few artists,
the great care and nicety of execution required, making
it a difficult art to attain, while the old preparation of
varnish colors placed difficulties in the way of even the
most practiced artists.

Water colors are now prepared expressly for painting
on glass, so that any one possessing a moderate knowledge of drawing, with some skill in the use of colors,
may succeed. None but transparent colors, or those
through which light is transmitted, can be used, making

the number of colors available for painting on glass necessarily limited. The most valuable for this purpose are the moist water colors procurable in metallic collapsible tubes : for yellow, Indian yellow, Italian pink, and yellow lake; for blue, Prussian blue and indigo; for red, madder lake, crimson lake, and scarlet lake ; for orange, burnt sienna; for brown, madder brown,Vandyke brown, sepia, and burnt umber; for black, India-ink and lamp-black; for purple, purple lake, or red and blue mixed; for green, mix yellow and blue; for scarlet, red and yellow. A white porcelain palette, free from specks and grit, is the best upon which to mix and arrange the colors. Use soft water for mixing the tints. For cake colors, use a weak gum water, taking care to have it quite dilute to prevent the colors cracking or peeling off; place each tint on a separate slab or saucer.

A suitable easel for holding the glass to be painted, is shown in the diagram (Fig. 25); this is a sloping frame, holding a sheet of glass, so arranged that it can be placed at any angle, and any convenient height for the artist.

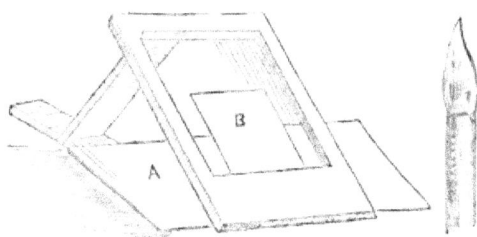

Fig. 25.

The glass or photograph to be painted, should be placed upon the clear glass in the frame (as shown at B). Upon the base board (A) is spread a sheet of pure white paper to reflect up the light through the painting; the light should fall on it from the left hand, and is best obtained

from a window facing the north sky. Glass has a smooth
and a rough side; the smooth side is that on which the
drawing is to be made; it may readily be determined by
drawing the finger-nail over the surface. The glass
should be carefully cleaned with water, to which a solution of ammonia has been added. A fine brush, or cheap
gold pen, may be used for drawing outlines, which should
be made with colors suited to the part of the painting
for which they are to be used; the foreground being
drawn with bolder lines than those more remote.

One of the most difficult things to accomplish in transparent painting on glass, perhaps, is to lay on a uniform
tint, free from lines or specks; as a clear blue sky without clouds. The brush should be well charged with the
blue tint, and the color spread or floated upon the glass
as evenly as possible, and afterwards equalized by a
careful application of the brush dabber : that is a camel-
hair brush cut down (as shown in Fig. 25), the edge of
which being afterwards passed through a flame so as to
remove any straggling hairs. The finger, also, may be
used as a dabber, and when used with dexterity, is very
effective. To take out the necessary lights, as those of
clouds, and to soften the edges, a stump made of leather
or paper may be used. In coloring photographs the outline and shading are provided; so that flat washes of
color are to be laid on, and then retouched and improved;
avoid covering the deepest shadows, thus destroying
their transparency. Breathe on it sufficiently to moisten
the colors, and carefully blend and harmonize the tints;
commence with the sky, then the middle distance should
be worked out, lastly the foreground. As the pictures
are necessarily small, a magnifying hand lens, such as
is used by artists for fine work, is desirable to assist one
in coming close to the lines with washes of color.

The brushes should be sable, of moderate size, and soft to the touch, and when charged with water, come to a good point without straggling hairs; some prefer a flat brush instead of a round one. It is well to have a sufficient number of brushes, and to use a different one for each tint. A piece of cloth should be used for cleaning brushes and dabbers, as neatness is very essential to success. An ordinary round-pointed pocket knife will be found useful for removing color. Etching-needles may be used for making minute touches of light, as on spears of grass; winter, snow, spring, and moonlight effects are produced chiefly by the skilful use of the knife and needle-points, to remove the color and produce strong white light in the picture. As pictures vary much in style, it would be difficult to give directions which would apply to all. Beginners should copy well-painted lantern slides at first, as this would guide in the colors to be used. Practice on waste pieces of glass and noting the effect in the lantern, would also prove beneficial and accustom the artist to regulate the tones of the picture in the best manner. When the picture is finished, it should be protected by a thin transparent varnish, such as photographers use, or a thin coat of Canada balsam. To prevent scratching, a glass, the same size as the picture, should be laid over it; and to prevent injuring from contact, a narrow rim of paper should be interposed between the glasses; they can then be bound or framed.

"Aniline colors have been used for photographic views with some success. They are brilliant and transparent, but require careful use to prevent the tints running one into the other."

Comic slides are often painted in a coarser manner, and oil paints are used. The method is very similar to that given for water colors: the same kinds of brushes, dabbers, and the same list of colors are used. The paints employed are sold in tubes; mastic varnish diluted with turpentine is used as a vehicle, sugar of lead as a drier. Comic or slip slides are generally painted on two pieces of glass, one of which is firmly fixed in the frame, the other movable; these glasses are so adjusted, that when the sliding glass is pulled out, an effect is produced which differs entirely from that shown when the glass is pushed in; as, for example, "The Windy Day;" the lady is seen passing along, fashionably dressed and equipped; the slip being drawn, she is shown in sad plight by the turned parasol, loss of false hair, bonnet, &c.; or a beautiful lily or tulip is seen; the slip is drawn, and a lovely fairy seems to float up from the flower. Chromatropes are constructed of two circular pieces of glass painted from the centre to the circumference of the circle with variously tinted rays and patterns, these are framed in brass frames, having grooves around them turned face to face, and when made to revolve reversely throw out beautiful and brilliant hues; according to the way in which they are made to turn, they expand or contract.

Statuary gives a much better effect, if the glass around it is covered with some opaque paint. Lampblack ground very fine with mastic varnish, a few drops of oil of cloves, and then brought to the right consistency with turpentine, is perhaps the best, as it does not rub off. "Opaque," an article manufactured by Mr. Gihon, of Philadelphia, is more easily applied, being used with water, and answers every purpose.

Figures which appear on the screen as black shadows, may be painted on the glass with these materials; or, to produce the same effect, designs may be cut from paper and pasted on the glass. Glass may be smoked or covered with opaque paint, and diagrams scratched upon it with a needle-point or sharp knife; the light passing through these lines appearing on the screen as a white chalk diagram on a blackboard. Still another way of preparing diagrams is to dissolve gelatine, such as is used in cooking; strain, and pour it over the glass, forming a thin film on its surface. When this is dry, the diagram is scratched on as before, and soft lead rubbed over the lines. Mottoes may be photographed on glass, and then colored, or the designs drawn with the pen or brush, and colored.

The Sciopticon is extremely well adapted for experiments and amusements, as its front lens can be drawn out, giving ample space for the introduction of figures and such like. Small china and wooden dolls, with but slight tissue-paper dress, may be made to twirl or move about in many curious ways; those with perfect faces are the best. They of course must be suspended by a silk or wire attached to the feet; but a hint is sufficient. Lizards, fish, and insects in the tank are always pleasing because they move. When one has but few slides, the entertainment may be varied by introducing some of the home-made objects, thus affording much amusement, with but slight expense and trouble.

CHAPTER VII.
CHEMICAL EXPERIMENTS.

CONTRIBUTED BY PROF. HENRY MORTON, Ph. D.

President of the Stevens Institute of Technology, Hoboken, N. J.

IN addition to the use of the magic lantern in its original office of exhibiting pictures, it will admit of a great variety of applications which enable the operator to produce countless variations in the effects developed, by which an endless variety and constant novelty can be secured.

For this purpose there is needed in the first placed the simple apparatus shown in our wood cut, consisting of a small tank, made by securing two plates of glass, about 4 x 5 inches, with four clamps, against a strip of rubber about ½ inch thick, bent into the three sides of a rectangle

Fig. 26.

and notched at the corners to facilitate its bending.

We then require one or more glass pipettes provided with elastic balls, such as are made by the rubber manufacturers. This little apparatus is shown in Fig. 27, where A is the rubber ball, B the glass globe of the pipette, and C its point drawn to a moderately fine orifice.

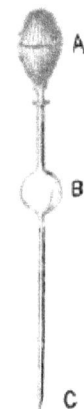

Fig. 27.

A few small pipettes made by simply drawing short pieces of glass tube to a fine point, are also useful.

In addition, a few bottles with such ordinary chemicals as will be mentioned further on, will complete the outfit.

Having placed the tank, three-quarters full of water, as an object in the lantern, a number of chemical reactions can be shown, as follows:

Experiment 1st. Pour in a little solution of sulphate

of copper, and mix it well with the water of the tank, then with the pipette run in, with more or less force, some diluted ammonia, pausing from time to time to observe the progress of the effect. On the screen will be observed the gathering of a tempest of black storm-clouds, which twirl around in violent commotion, as if urged by a tornado of wind, but as the action continues, these clouds will melt away, and leave the entire field of a serene and beautiful sky-blue.

By now throwing in some diluted sulphuric acid, the same changes can be reproduced, and so on alternately for a number of times. Then when the tank is clear, with an excess of acid, let fall a few drops of a solution of ferrocyanide of potassium from a small pipette, and rich red curdled clouds of ferrocyanide of copper will form with a beautiful appearance.

Experiment 2d. Having rinsed the tank, or taken a fresh one with water in it as before, add to this some solution of litmus, until the whole acquires a purplish-blue tint. Now throw in very gently a little very dilute acid, and allow it to diffuse. On the screen will appear the image of a beautiful sunset sky, with its changing tints of drifting clouds.

When all has changed to red, add ammonia, and so reverse the change, which may then be repeated.

Experiment 3d. Proceed exactly as in the last case, but with a solution of cochineal in place of litmus. The red color will then be changed by the acid to a brilliant yellow, and by ammonia to a rich purple.

Experiment 4th. Into a tank of water drop slowly a strong solution of the acid perchloride of tin. This on the screen will resemble the eruption of a submarine volcano.

When a pretty strong solution has thus been made in

the tank, put in it a strip of sheet zinc, and long leaf-
like blades of metallic tin will at once be seen to shoot
out in all directions.

Experiment 5th. Make a concentrated solution of
crystals of urea in alcohol of about 95 per cent. (The
common 85 per cent. alcohol will not answer.) Let a
few drops of this fall on a glass plate, and with the
finger spread it rapidly over the surface, and then at
once place it as an object in the lantern. After about a
minute, blow gently on the plate with a bellows (not
with the breath), and at once on the screen will be seen
the growth as of frost crystals shooting over the field
in all directions.

Experiment 6th. If sulphate of copper in solution is
mixed with enough gum-arabic water to make the solution form a continuous film, when flowed like collodion
on a clean glass, and such plates are allowed to dry
slowly in a nearly horizontal position, a very beautiful
crystalline vegetation will set in, which varies in its
character with the proportion of gum used, and will
make objects well fitted for exhibition with the lantern.

In place of sulphate of copper, we may use nitre, or
ferrocyanide of potassium, with the production of an
entirely new class of forms.

By placing the plates so covered with crystals over a
leaden dish, in which is a little fluor-spar, moistened
with sulphuric acid, and warmed slightly (giving off
fumes of hydrofluoric acid), permanent etchings may be
prepared, which are also very beautiful objects for the
lantern.

These are only a few of the experiments of this character which can be performed with the lantern, but they
will indicate the direction in which each one can be a
discoverer and inventor for himself.

MISCELLANEOUS EXPERIMENTS.

THE SCIOPTICON TANK (Fig. 28) is free from projecting clamps and so passes freely upon the stage in front of the condenser. It serves as a dry cage for insects, &c., a cell to show liquids and life in water, a tank for the exhibition of chemical reactions, and with wires protected and bent over the ends it can be used in connection with a galvanic battery. It is the most convenient for the preceding experiments, as well as for these which follow.

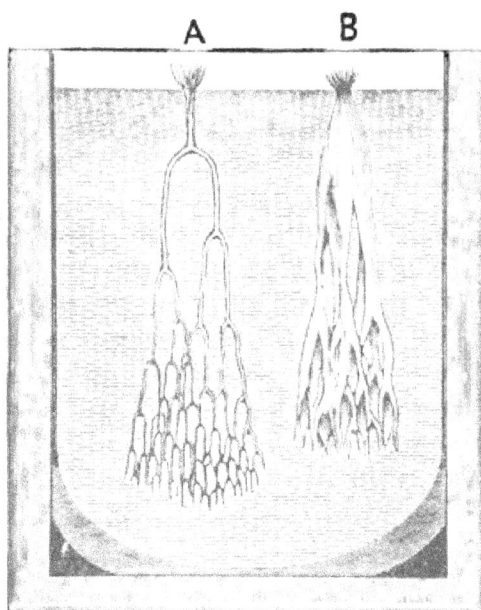

Fig. 28.

COHESION FIGURES.—The cohesion figures known as Tomlinson's are both interesting and beautiful, and can be shown as follows: Fill the tank to within half an inch of the top with alcohol and slide it into place upon

the stage; now with a glass rod, or small brush, dipped
in any of Judson's aniline dyes, touch the side of the
tank gently, so as to leave a drop on it. This drop, directly as it touches the alcohol, will go straight down for
half an inch or so, and then break out into two branches;
these again will break in four, and so on, until by the
time the dye gets to the bottom of the tank it will have
formed some hundreds of delicate branches. As this
action is reversed on the screen, the branches appearing to shoot upwards, the effect is much heightened.
A (Fig. 28), shows the form assumed. By placing at in-
tervals of half an inch drops of different colors, as their
branches commingle, the effect reminds one of a shower
of different colored rockets. If we now take another
tank, and fill it with coal oil, and put a drop of fusel
oil into it, we get an entirely different figure, as shown
at B. The fusel oil is best colored.

CAPILLARY ATTRACTION can be strikingly shown to a
large audience. A series of glass tubes of different sizes
are fitted into a piece of wood which rests on the top of
the tank, and dips down to near the bottom; when the
tank is filled with water, which is best tinted, the different heights of the water, according to the fineness
of the tubes, will be shown clearly on the screen. The
curve shown by the liquid rising between two pieces of
glass can be shown in the same manner, the colored
water forming a pretty gradation of color between the
highest and lowest part.

CRYSTALLIZATION.—By filling the tank with a saturated solution of Glauber's salts, and allowing it to cool,
it will appear transparent on the screen, but by dropping
one small crystal into it the whole mass will be seen to
shoot out into beautiful crystals.

The crystallization of many other substances, such as bichromate of potash, alum, &c., and the precipitation of iodides of silver, mercury, and other salts, all form beautiful objects on the screen.

THE DEVELOPMENT OF A PHOTOGRAPH ON THE SCREEN. —For this we require a tank with one of its faces of yellow glass, which side should be next the condenser. Place a small statuette in the rays of the lantern, and having prepared a small plate with collodion and sensitized it, expose in the camera for about a minute; then, having filled the trough with developing solution, place in it the slide, and as the development proceeds the image will gradually appear on the screen. A transparency might then be made from this, and, after drying, shown on the screen, thus illustrating the formation of a photographic lantern slide.

CHANGING COLORS.—A glass coated with a mixture of gelatine and chloride of cobalt, when placed in front of a slide, will give a rosy effect to the picture, which, however, from the effect of the warmth of the lantern, will gradually change to purple and then to blue. On becoming damp again it will resume its red color, and can be used over and over again.

COMPLEMENTARY COLORS.—A number of beautiful effects, showing complementary colors, may be obtained with the Sciopticon. If we insert a piece of green glass, having any design cut out of black paper and pasted on it, we shall see on the screen a black design on a green ground; but by bringing another light into the room or turning up the gas, the black design will at once appear to the eye as a brilliant pink.

By making apertures in a card slide, as circles, squares,

7

or diamonds, say a fourth of an inch in diameter, and
covering them with bits of colored gelatine, or by simply
using the tinters of the Sciopticon, many curious effects
in complementary colors may be obtained.

FAIRY FOUNTAIN.—The effect of what is known as the
"Fairy Fountain" can be prettily illustrated in the following manner : A small table fountain is placed at a
distance of about four feet in front of the lantern ; by
curtains or otherwise the lantern is then hidden from
the spectators, so that they see only the fountain illuminated by the rays coming from the lantern. When the
fountain is made to play, every drop seems transformed
into a diamond, and by passing colored glass in front of
the lantern the effect is striking and beautiful ; but when
the rays from a bisulphide of carbon prism are allowed
to fall on it, then is the best effect produced.

THE RAINBOW.—A card with a curved slit, one-six-
teenth of an inch (Fig. 29), will throw on the screen a
simple semicircle of white
light ; but when a prism
is held in front of the objective, the bow at once
assumes all the natural
colors of the rainbow. As
the direction of the rays
is changed, the range of
the instrument has to be
elevated, to bring the bow upon the screen. By using
two lanterns, projecting a view with one and the bow
with the other, a very natural effect may be produced.

Fig. 29.

A MAGNET AND IRON FILINGS.—Fix a small magnet
to a glass slide, and carefully arrange a funnel opening

above the poles in the lantern; then allow iron filings to fall gently down the funnel, which will appear like large blocks attracted upward by a huge magnet.

ASTRONOMICAL CARDS.—The cards may be cut to the size of the crystal slide, that is 3¼ by 4¼ inches, so as to be used in the grooved frame, like an ordinary glass slide. After correctly dotting a constellation of stars (which may be done by the use of theorem paper and a good map of the heavens), pierce the card at the several points, say with a darning needle, which may be made to show stars of different magnitudes by gauging the depth of the insertion.

To illustrate the Solar System, punches of different sizes might be used and bits of colored gelatine, covering the aperture, might indicate the tints attributed to each member.

PINHOLE OUTLINES.—Cards in shape of glass slides and just thick enough to be sufficiently stiff, may be pricked to show maps, mottoes, figures, diagrams, or any simple illustration. They require but little skill and show very distinctly.

PERFORATIONS.—Two pieces of perforated paper or tin made to slide little by little over each other, in front of the condenser, and modified more or less by the tinters, produce beautiful symmetrical forms in great variety.

PERSISTENCE OF VISION.—Apertures, as in a paper card, when moved rapidly in all directions in the plane of the slide, appear as lines of light on the same principle that a lighted stick waved about produces lines of light. A new slide, called the kaleidotrope, is constructed and hung to exhibit this curious effect.

THE PHOTODROME.—The photodrome, as shown at the Polytechnic, may be made at a very small expense. To produce this effect we require a rapidly moving disk (having one or more slits cut in it) revolving in the place where the slide is placed, and also a larger one placed at some distance—the latter representing a wheel, the spokes of which are painted in black on a sheet of white cardboard. When this is made to revolve rapidly in the rays coming from the lantern, all trace of the spokes will be completely lost; but on causing the small disk to revolve at nearly the same speed as the larger, the latter will appear to be moving slowly, although moving rapidly, and by increasing the speed of the smaller wheel, the larger will gradually appear to slacken in speed until it appears to be motionless, and then apparently begin to move in an opposite direction to which it is really revolving.

SILHOUETTES, &c.—Paper patterns, silhouettes, &c., suspended by a thread attached to the feet, and twirled before the condenser, give a very amusing and curious effect.

GALVANIC ACTION.—Fill the tank with a solution of nitrate of silver, and introduce at each end two wires from a small battery; from one of the wires a beautiful silver tree will immediately begin to grow. The experiment may be varied by substituting acetate of lead for a lead tree.

Litmus solution, neutralized, will gradually redden around one point, while around the other it will assume a blue tint.

With a solution of cochineal, the red color will be changed by the acid to a brilliant yellow, and by the ammonia to a rich purple.

NATURAL OBJECTS, as leaves, plants, fibres, texture of cloth, thin sections of wood, bone, &c., appear in distinct outline upon a white ground. Live animals in the tank, as insects, larvæ of gnats, shrimps, worms, lizards, &c., appear as huge monsters upon the screen, and excite a lively interest by their eccentric movements.

VERTICAL LANTERN.—Some very interesting experiments require the slides to lie in a horizontal position. This is commonly effected by reflecting the light up through the glass plate and the objective lens, and then by another mirror reflecting the image horizontally to the screen. A lantern appendage of this sort in now in the trade, at $20.

But this is equivalent to placing the slide at least four inches from the face of the condenser, which, at best, puts it at great disadvantage, and then there is the loss of light by two reflections.

The oil light cannot well bear these drawbacks, the lime light is better; but with the lime light the Sciopticon may be placed on end, as shown at Fig. 30.

When attached to its carrying box, in the ordinary way, it may be held in this position over the edge of a table, so as to be conveniently operated. The front flame-chamber glass will protect the condenser from its greater liability to become heated.

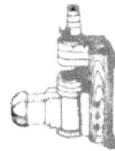
Fig. 30.

A glass disk, clean cut, and slightly larger than the condenser, answers for the slide plate; and if a rubber band be stretched about its periphery, like the tire of a wagon wheel, it will become a tank for fluids.

With this arrangement, a mirror at an angle of 45° above the objective will throw the effect upon the screen without appreciable loss of light.

RIPPLE WAVES.—Fill the tank, as it rests on the vertical lantern, with clear water, when taps on the edge of the glass will start ripple waves, which will be seen on the screen in varied harmonious arrangements of form. Touching the surface with the point of a fine wire will start the waves in circles. Vibrations effected by drawing a fiddle bow across the edge are seen to vary according to the different tones produced.

ADHESION FIGURES.—Drops of various oils upon the surface of the water, essential oils for instance, will exhibit various interesting adhesion figures, each oil assuming some peculiar form of outline.

MAGNETIC CURVES.—A thin bit of magnetic steel, say three-fourths of an inch long by one-eighth wide, cemented on the under side of a glass plate, will attract fine iron filings scattered upon the plate into curves, illustrating the deviation of the magnetic attraction at either pole and the neutral axis in the centre of the magnet. A few taps on the glass will assist the arrangement.

CHAPTER VIII.

Descriptive Lectures.

CONCERT EXERCISES.

THE value of visible illustrations as a means of imparting instruction, and of affording rational entertainment, depends much on the accompanying oral explanations.

Except to a very limited extent, it is not practicable (as many seem to suppose it is), to forward with a miscellaneous selection of magic lantern slides a printed lecture.

In the absence of special provisions for supplying this demand, some general hints in this direction may here prove acceptable.

In some assemblages (possibly in some Sunday-schools), very little can be said to advantage on account of the prevailing noise and confusion. The exhibitor having (for love or money) accepted the situation, the question arises as to how to make the best of it.

In such cases in particular it is politic, as well as proper, to select slides unexceptionable in their influence. Grotesque and ridiculous representations gratify a depraved taste, and render a demoralized company still more unruly. It is better to please by what is strikingly excellent and beautiful.

Without assuming the attitude of a reformer, one may take advantage of the lull of expectancy preceding a change of scene to give in a natural voice some interesting particulars of the forthcoming picture.

> " Your mystical lore,
> As coming events cast their shadows before,"

will be respected, and you may be able, by judicious management, to strengthen your position on vantage ground. Even in a civilized assembly (and we may well hope to find ourselves in no other), some tact is needful, as well as agreeable speech and faultless manipulation.

BIBLE PICTURES.

Among standard colored lantern slides, Bible pictures properly take the lead. They embody the genius of the

most gifted artists, in connection with subjects of the
most thrilling interest to mankind.

We may name the picture, particularizing when necessary its several parts, and then repeat the Scripture
which is illustrated.

Take, for example, Adam and Eve in Paradise; the
luxuriant foliage, the lion, the ox, the horse, the birds,
and alas! the subtle serpent.

" In the beginning God created the heaven and the earth.

" And God said, Let us make man in our image, after our likeness; and let them have dominion over the fish of the sea, and over
the fowls of the air, and over the cattle, and over all the earth, and
over every creeping thing that creepeth upon the earth.

" So God created man in his own image; in the image of God
created he him; male and female created he them.

" And the Lord God planted a garden eastward in Eden; and
there he put the man whom he had formed."—Gen. 1: 1, 26, 27;
2: 8.

Or take the scene where Joseph presents his father
to Pharaoh. Mark the postures of each, and consider
the manners of the times.

" And Joseph brought in Jacob his father, and set him before
Pharaoh; and Jacob blessed Pharaoh. And Pharaoh said unto
Jacob, How old art thou? And Jacob said unto Pharaoh, The
days of the years of my pilgrimage are a hundred and thirty years:
few and evil have the days of the years of my life been, and have
not attained unto the days of the years of the life of my fathers in
the days of their pilgrimage. And Jacob blessed Pharaoh, and
went out from before Pharaoh."—Gen. 47: 7, 8, 9, 10.

Thus Scripture, to any desired extent, may be readily
selected appropriate to any Bible picture, from Adam
and Eve in Eden to St. John's vision of the Celestial City.
So the exhibitor has ample material at hand for shaping
an effective and charming discourse, suited to any series
of Bible pictures which he may have to show.

The Bible is, par excellence, the storehouse of unfailing supplies for the

SUNDAY-SCHOOL.

In this modern institution, as elsewhere, there are many duties to be performed, and more ways than one of doing each of them. We will indicate, in this connection, one way of using the Sciopticon. Each member of the school takes a small moneyed interest in the concern at the outset, which insures his taking a more lively interest in the success of the enterprise afterwards.

The apparatus is strictly in the hands of an authorized keeper, because lax regulations suppress all genuine enthusiasm.

The operator arranges his slides in proper order and position, and so is able to avoid ridiculous blunders. His characters are introduced on time, steady and upright, and his scenery glides into place as if seen from the deck of a moving steamer.

It is good policy to enlist as many pupils as possible into active service, thus incidentally enlisting the sympathies of as many circles of relatives and friends.

Suppose repentance is the theme, and the " Prodigal's Return " is illustrated upon the screen. A pupil, fully prepared, stands in his place and recites the whole parable as found in Luke 15.

Another pupil, rising in his class, recites:

" Therefore also now saith the Lord, Turn ye even to me with all your heart, and with fasting, and with weeping, and with mourning. And rend your heart, and not your garments, and turn unto the Lord your God ; for he is gracious and merciful, slow to anger, and of great kindness, and repenteth him of the evil."—Joel 2: 12, 13.

A third voice rings out clearly :

" Let the wicked forsake his way, and the unrighteous man his

thoughts; and let him return unto the Lord, and he will have mercy
upon him; and to our God, for he will abundantly pardon."—
Is. 55: 7.

Passages bearing **on** repentance and forgiveness **are
very numerous,** from which selections can be made **to
any extent desired.** Illustrations with fewer **relations
to parallel passages may** be coupled with **others to extend the exercise to** proper length.

Selections also from modern writers, well rendered,
give pleasing variety and artistic effect to the performance. The sacred poems of N. P. Willis, for example,
are very appropriate. **The following** extracts may serve
as specimens :

ABRAHAM'S SACRIFICE.

.   .   .   .   He rose up, and laid
The wood upon the altar.   All was done.
He stood a moment, and a deep, quick flush
Passed o'er his countenance; and then he nerved
His spirit with a bitter strength, and spoke—
"Isaac! my only son!"   The boy looked up.
"Where is the lamb, my father?"   Oh, the tones,
The sweet, familiar voice of a loved child!
What would its music seem at such an hour?
It was the last deep struggle.   Abraham held
His loved, his beautiful, his only son,
And lifted up his arm, and called on God,
And lo! God's angel stayed him—and he fell
Upon his face, and wept.

HEALING OF THE DAUGHTER OF JAIRUS.

.   .   .   .   The Saviour raised
Her hand from off her bosom, and spread out
The snowy fingers in his palm, and said—
"Maiden! arise!"—and suddenly a flush
Shot o'er her forehead, and along her lips,

And through her cheek the rallied color ran ;
And the still outline of her graceful form
Stirred in the linen vesture ; and she clasped
The Saviour's hand, and fixing her dark eyes
Full on his beaming countenance, AROSE!

CHRIST WEEPING OVER JERUSALEM.

. . . How oft, Jerusalem! would I
Have gathered you, as gathereth a hen
Her brood beneath her wings, but ye would not!

He thought not of the death that he would die—
He thought not of the thorns he knew must pierce
His forehead—of the buffet on the cheek—
The scourge, the mocking homage, the foul scorn !
Gethsemane stood out beneath his eye
Clear in the morning sun, and there he knew
While they who "could not watch with him one hour"
Were sleeping, he should sweat great drops of blood,
Praying the "cup might pass." And Golgotha
Stood bare and desert by the city wall,
And in its midst, to his prophetic eye,
Rose the rough cross, and its keen agonies
Were numbered all—the nails were in his feet—
The insulting sponge was pressing on his lips—
The blood and water gushing from his side—
The dizzy faintness swimming in his brain—
And, while his own disciples fled in fear,
A world's death-agonies all mixed in his!
Ay—he forgot all this. He only saw
Jerusalem—the chosen—the loved—the lost !
He only felt that for her sake his life
Was vainly given, and, in his pitying love,
The sufferings that would clothe the heavens in black
Were quite forgotten. Was there ever love,
In earth or heaven, equal unto this?

Longer or shorter extracts may be used as occasion
requires. The following are titles of others, equally

beautiful, and descriptive of subjects illustrated by lantern slides: "Hagar in the Wilderness," "The Shunamite," "Jepthah's Daughter," "Hannah and Samuel," "Absalom," "Rispah with her Sons," "Baptism of Christ," "The Widow of Nain," "The Raising of Lazarus," "Christ's Entrance into Jerusalem," and "Scene in Gethsemane."

The following poem, by an author unknown to us, will be inserted entire, as it so vividly portrays the mind of the parent and the love of the Saviour for children, and so graphically describes the picture of "Christ Blessing Little Children:"

"The Master has come over Jordan,"
　Said Hannah, the mother, one day;
"Is healing the people who throng Him,
With a touch of his finger, they say.

"And now I shall carry the children,
　Little Rachel, and Samuel, and John;
I shall carry the baby Esther,
　For the Lord to look upon."

The father looked at her kindly,
　But he shook his head, and smiled;—
"Now, who but a doting mother
　Would think of a thing so wild?

"If the children were tortured by demons,
　Or dying of fever, 'twere well;
Or had they the taint of the leper,
　Like many in Israel."

"Nay, do not hinder me, Nathan,
　I feel such a burden of care,
If I carry it to the Master,
　Perhaps I shall leave it there.

"If He lay His hand on the children,
　My heart will be lighter, I know,
For a blessing forever and ever
　　Will follow them as they go."

So over the hills of Judah,
　Along by the vine-rows green,
With Esther asleep on her bosom,
　And Rachel her brothers between;

'Mong the people who hung on His teaching,
　Or waited His touch and His word,
Through the rows of proud Pharisees listening,
　She pressed to the feet of the Lord.

"Now why shouldst thou hinder the Master,"
　Said Peter, "with children like these?
Seest not how from morning till evening
　He teacheth, and healeth disease?"

Then Christ said, "Forbid not the children:
　Permit them to come unto Me,"
And He took in His arms little Esther,
　And Rachel He set on His knee.

And the heavy heart of the mother
　Was lifted all earth-care above,
As he laid His hand on the brothers,
　And blessed them with tenderest love.

And He said of the babe in His bosom,
　"Of such is the kingdom of heaven,"—
And strength for all duty and trial,
　That hour to her spirit was given.

A little poem published by the American Tract Society,
called the "Old, Old Story," could be used in connection
with a series of six slides.

The "Song of the Pilgrimage," and "Christiana and

her Children," are much used in connection with the
corresponding slides. These published exercises afford
practical hints, applicable also to Bible slides.

Singing should be introduced at every convenient
opportunity, not only for its general good effect, but
that each individual may participate directly in the
exercises.

Texts of Scripture, and other selections, recited in
this way at the rehearsals, and at the concert, become
fixed in the memory of all. Who cannot remember
such recitations heard in childhood, even to the tones
and inflections of the voice—of voices, maybe—not now
heard among the living?

These modest recitations require no parade upon an
illuminated rostrum; an occasional omission is not very
noticeable. The exercises can be arranged by the
superintendent, divided among the teachers, assigned
to the pupils, and committed to memory by them without severe labor on the part of any.

One or two slides for the concert exercise, with, say
a dozen or so for subsequent recreation, answers the
purpose. Such a concert exercise, well gotten up, may
be several times repeated with growing interest.

It often occurs in schools, where the burdens and
duties are monopolized by the few, that the many become impatient of control and hard to please. A hundred pictures in such cases hardly suffices, and a repeti-
tion of the same is scarcely tolerated.

An earnest worker in the Sunday-school, therefore,
can accomplish more good, not by trying to do everything himself, but by skilfully assigning work for others,
and seeing that it is properly done. After all, there
will be enough left for pastor and superintendent to do
and say, especially when it comes to slides selected from

Class III or V of the appended catalogue, which will
require a lecturer well informed in relation to

BIBLE LANDS.

The following descriptions are selected from the
" Bible Dictionary," " Bible Lands," " The Land and the
Book," " Bayard Taylor's Travels," &c., to suit the slides
in Class III.

As works on Egypt are less common than the Bible
Dictionary, a description of each of the twenty Egyptian
views is given.

JERUSALEM.

(For description of the City, and view from Mount of Olives, see Catalogue, Class III.)

THE TEMPLE AREA.—The Temple Area, the precincts
known to Christians as the Mosque of Omar, but called
by the Moslems the "Dome of the Rock," the harem
more sacred to Moslems than any spot on earth, except
Mecca, is jealously guarded by the Turks. It contains about thirty-five acres, a large portion of which
is sprinkled with pomegranates and cypresses, with here
and there a shrine. Above this space rises the platform
of the great mosque, paved with marble, and ascended
by a flight of white marble steps, surmounted by a beautifully carved screen or open gateway, also of white
marble. The edifice is an octagon of about one hundred
and seventy feet diameter. There are four doors at the
opposite cardinal points. The dome is sustained by four
great piers, and has twelve arches, which rest on columns.
The mosque is very beautiful with a kind of Moorish
beauty. The octagonal walls below the dome are covered with porcelain mosaic; the roof inside is of the
richest woods, inlaid and carved; the floors of marble

mosaic; the windows like jewelry, of small pieces of
Venetian stained glass. Beautiful columns, and an
elaborately worked balustrade, surround the holy stone
(Es Sakrah, the rock), which Moslems believe to be the
centre of the world, suspended from heaven by an invisible golden chain. It is a mass of the native rock of
Moriah, the sloping summit or peak of the hill; all the
rest of the ridge was cut away when levelling the platform for the temple and its courts.

THE TOWER OF HIPPICUS.—The only castle of any
particular importance is that at the Jaffa Gate, commonly called the "Tower of David." The lower part
is built of huge stones, roughly cut, and with a deep
bevel around the edges. It is believed by many to be
the Hippicus of Josephus, and to this idea owes its chief
importance, for the historian makes that the point of
departure in laying down the line of the ancient walls
of Jerusalem.

THE CHURCH OF THE HOLY SEPULCHRE.—The Church
of the Holy Sepulchre is now in the joint possession of
all the Eastern Christian sects. Greeks, Latins, Arme-
nians, and Copts have each a chapel within its inclos-
ures, which embrace the alleged sites of the place of the
crucifixion and the tomb of the Redeemer. It has been
built at many different periods, and under various circumstances.

"The front is a fine specimen," says Lord Nugent,
"of what is called the later Byzantine style of architecture." As lately as 1808, the whole of the principal
cupola, and a great part of the church, were destroyed
by fire. But some parts, and especially the Greek chapel,
occupying the whole of the eastern end of the nave,
have been restored with good taste and judgment, and

are magnificent in their proportions and decorations. The sepulchre looks very much like a small marble house. It stands quite alone, directly under the aperture in the centre of the dome.

THE JEWS' PLACE OF WAILING.—No sight meets the eye in Jerusalem more sadly suggestive than the wailing-place of the Jews, in the Tyropean, at the base of the wall which supports the west side of the Temple Area, where some ancient stones still mark the old walls of the temple. In past ages the Jews have paid immense sums to their oppressors for the miserable satisfaction of kissing these stones, and pouring out lamentations at the foot of their ancient sanctuary. With trembling lips and tearful eyes they sing: "Be not wroth very sore, O Lord, neither remember iniquity forever; behold, see, we beseech thee, we are all thy people. Thy holy cities are a wilderness; Jerusalem is a desolation. Our holy and beautiful house, where our fathers praised thee, is burned up with fire, and all our pleasant things are laid waste."

THE GOLDEN GATE AT JERUSALEM.—In former days the gates of towns were of the utmost importance; they were the means of ingress and egress, and usually had rooms over them, and, above these, watch-towers, so that the approach of an enemy might be seen beforehand. The Golden Gate, in the east wall of the Temple Area, is ancient, and the interior of it ornamented with rich and elaborate carving in good Grecian style. It is now walled up.

GARDEN OF GETHSEMANE.—"Then cometh Jesus to a place called Gethsemane, and saith unto the disciples, Sit ye here, while I go and pray yonder."—Matt. 26: 36.

8

Across the brook Kedron, probably at the foot of Mount Olivet, was the "place" or "farm" of Gethsemane. There seems to have been a garden, or rather orchard, attached to it, and to its grateful shade we read that our Lord often resorted with His disciples. At present a modern garden marks the site of the ancient one with eight venerable olive trees, which some claim grew there in the Saviour's time. It has been argued that Titus cut down all the trees about Jerusalem. The probability would seem to be that they were planted by Christian hands to mark the spot; unless, like the sacred olive of the Acropolis, they may have reproduced themselves.

BETHLEHEM.—Bethlehem was in existence when Jacob returned from his long sojourn in Padan Aram. Here Rachel died. It was in the neighboring fields, in later times, that Ruth, the Moabitess, went gleaning when she came with her mother-in-law, Naomi, to dwell in the land of Israel. It was the birthplace of David, but is best known to us as the birthplace of the Redeemer, great David's greater son and Lord. "On the plains near were the shepherds abiding in the fields, and keeping watch over their flocks by night, when lo! the angel of the Lord came upon them, and the glory of the Lord shone round about them, and they were sore afraid. And the angel said unto them, Fear not, for behold, I bring you good tidings of great joy, which shall be to all people; for unto you is born this day, in the city of David, a Saviour which is Christ the Lord."—Luke 2: 8–14.

HEBRON.—Hebron is one of the most ancient cities in the world still existing. "It was built," says a sacred writer, "seven years before Zoan in Egypt."–Num. 13: 22, and was a well-known town when Abraham entered

Canaan 3780 years ago. Sarah died at Hebron, and Abraham then bought from Ephron, the Hittite, the cave of Machpelah, to serve as a family tomb. Jacob gave commandment to his sons, "Bury me with my fathers in the cave that is in the field of Ephron, the Hittite. There they buried Abraham and Sarah his wife. There they buried Isaac and Rebekah his wife, and there I buried Leah." And his sons did unto him according as he commanded them, and buried him in the cave of Machpelah. The massive walls of the harem or mosque, within which the cave lies, forms the most remarkable object in the whole city. Hebron now contains about 5000 inhabitants, of whom some fifty families are Jews. It is picturesquely situated in a narrow valley, surrounded by rocky hills.

"THE POOL OF SILOAM" is one of the few undisputed localities in Jerusalem, still retaining its old name. It is of no considerable size, being eighteen feet broad and nineteen deep. It is, however, never full, having in it usually about four feet of water. It is a complete ruin. It was to this pool that our Lord sent the blind man, after he had anointed his eyes with clay. It was to Siloam that the Levite was sent with the golden pitcher on the last day of the feast of Tabernacles, and from it he brought the water which was then poured over the sacrifice, in remembrance of the water that flowed from the rock Rephidim.

GENESARET, OR SEA OF GALILEE.—This view exhibits a portion of that large inland sea through which the Jordan flows from north to south. It is some thirteen miles long and six broad, and is remarkable for the lowness of the basin in which it lies, being about seven hundred feet below the level of the ocean. No less than nine cities stood on the very shores of the lake. A great part of our Lord's life was spent near it. Here he taught the people out of Peter's ship, and wondrously filled the nets, so that they brake; walked on the waves, rebuked the winds, and calmed the sea. From the castle Saphet a vast panorama, embracing a thousand points of historic and sacred interest, is presented to the eye. Saphet is truly a high tower. Here are beveled stones, as heavy and as ancient in appearance as any ruins in the country, and they prove that this has been a place of importance from a remote age.

BATHS AND CITY OF TIBERIAS.—The sea of Galilee is also called the sea of Tiberias, from the celebrated city of that name. About a mile south from the original site of the city, along the shores, are the celebrated warm baths, which the Roman naturalists reckoned as among the greatest known curiosities of the world. The water of these springs has a sulphurous and most disagreeable smell, and is so nauseous that it cannot be drank, and is not used internally. The baths, however, have a great medicinal reputation. There is but one common bathing cistern, where the water is hot enough to cook an egg—from 130° to 140° Fahrenheit—yet it is always crowded with the lame, the halt, the withered, and the leprous.

NAZARETH.—Nazareth is situated among the hills which constitute the south ridges of Lebanon, just before they sink into the Plain of Esdraelon. It derives its celebrity from its connection with the history of Christ. The "Fountain of the Virgin" is situated at the northeastern extremity of the town. The brow of the hill is still called the Mount of the Precipitation (Luke 14 : 29), and is half a league southward of Nazareth.

THE VALLEY OF JEHOSHAPHAT.—The Valley of Jehoshaphat was the favorite burying-place of the Jews from the earliest times; accordingly we find in it a number of remarkable tombs. The monolith of Zachariah is a cubical block of about twenty feet every way, and surmounted by a flattened pyramid of at least ten feet elevation. It is one solid mass hewn out of the mountain, the adjacent rock being cut away, so that it stands entirely detached; there is no known entrance. The tomb of St. James shows a fine front to the west. The cave extends forty or fifty feet back into the mountain. Some two hundred feet north of this is the tomb of Absalom. The entire height of this very striking "pillar" cannot be less than forty feet. Believing it to be Absalom's tomb, the natives throw stones against it, and spit at it as they pass by. Close to this monument, on the northeast, is the reputed tomb of Jehoshaphat.

"THE DEAD SEA," says Dr. Thomson, "without any reference to what others have said, I can testify to the following facts: The water is perfectly clear and transparent. The taste is bitter and salt, far beyond that of the ocean. It acts upon the tongue and mouth like alum, smarts in the eyes like camphor, produces a burning, pricking sensation, and it stiffens the hair of the head much like pomatum. The water has a much greater specific gravity than the human body, and hence I did not sink lower than to the arms when standing perpendicularly in it. We saw no fish nor living animals in the water, though birds were flying over it unharmed. All of us noticed an unnatural gloom, not upon the sea only, but also over the whole plain below Jericho. It had the appearance of Indian summer in America, and like a vast funeral pall let

down from heaven, it hung heavily over the lifeless
bosom of this mysterious lake." Its area is about two
hundred and fifty square geographical miles. At its
northern end it receives the stream of the Jordan. The
depression of its surface, and the depth which it attains
below that surface, combined with the absence of any
outlet, render it one of the most remarkable spots on
the globe.

THE FORDS OF THE JORDAN.—The reach of the Jordan here shown is the place to which pilgrims of the
Greek Church resort every year, in Holy Week, to renew
their baptism by bathing in the Jordan, and it is the
spot which tradition points out as the place where our
Saviour was baptized. The Jordan is a rapid and tor-
tuous stream, interrupted by many rapids, and annually
" overflows his banks all the time of harvest." So far
as this overflow extends there is a belt of luxurious
vegetation, but beyond it the ground is barren.

EGYPT.

From time immemorial Egypt has been an object of
interest to the rest of the world. Almost the dawn of
Scripture light breaks upon the rocks and sands of this
wonderful valley, whose vast river diffuses fertility
wherever it flows. Here the children of Israel served
the Pharaohs four hundred and thirty years and grew
into a great nation. From the banks of the Nile they
set out on that marvelous pilgrimage to Sinai and Zion,
those two rocky pinnacles whence the splendors of the
Law, and the mild and beneficent radiance of the Gospel,
beamed forth upon mankind.

A TRAVELER'S NILE BOAT, OR "DAHABEEK."—The traveler who visits Egypt can avail himself of public conveyance as far as Cairo, but if he desire to visit the remains of ancient grandeur that lie to the south, he must engage a Nile boat, which becomes, for the time being, both the means of locomotion and his home; and as all the points of interest are near the river, a more commodious plan for visiting them could hardly be devised. As there are no towns above Cairo everything in the shape of comforts and luxuries must be provided before setting out.

STREET IN CAIRO.—The streets in Cairo, like those of most Oriental towns, are narrow, being some eight or ten feet wide. The houses are mostly three stories in height, each story projecting over the other, and the plain stone walls are either whitewashed or striped with horizontal red bars, as seen in the picture.

The beautiful latticed windows, "masharobeahs," are the chief ornament of the old Mameluke houses in Cairo. The wood seems rather woven in the loom than cut with the saw and chisel.· Through these lattices of fine network, with borders worked in lace-like patterns, and sometimes tipped with slender turrets, the Cairo ladies sit and watch the crowd passing to and fro, themselves unseen. "The mother of Sisera looked out at a window and cried through the lattice, Why is his chariot so long in coming?"—Jud. 5 : 28. Donkey-riding in the streets, and bazars, is almost universal. The animals are small but strong. The driver runs behind, gives the donkey a punch, cries "O man, take care! O boy, get out of the way!" and the rider is hurried into a confusion of other donkeys, loaded camels, water-carriers, and footmen. To one unaccustomed to donkey-riding it seems

as hazardous as going on foot.   The streets of Cairo are
watered several times a day, and are nearly always cool
and free from dust.

FERRY AT OLD CAIRO.—Old Cairo is situated about two
miles from modern Cairo.   The wonderful clearness and
brilliancy **of the Eastern** atmosphere; the absence of
smoke, **charcoal alone being** burned; the picturesque
effect of the ruin into which many of its great monuments are falling; the rich, green valley of the Nile; **the**
river; the Pyramids in the distance; and **the** fading
of the landscape into the boundless haze of the Lybian
desert, constitutes a scene which, for splendor and interest, is perhaps unequaled in the world.   The taste for
gaudy and fantastic coloring has been for ages a distin-
guishing feature of Eastern embellishment.   The alternate red and white stripe is conspicuous on the sails of
the ferry boats, which are constantly passing back and
forth between Cairo and the island of Rhoda opposite
Here we have a group of Arabs from the desert, with
their camels, dealers in oranges, vegetables, sugar-cane,
&c.   For picturesqueness of costume, there is nothing
like the East; the flow of the drapery so simple and
natural, the coloring so deep and brilliant.

TOMBS OF THE MEMLOOK KINGS AT CAIRO.—These
tombs are fine specimens of Saracenic architecture, and
were erected in the thirteenth and fourteenth centuries.

PYRAMIDS.—The Pyramids of Gizeh, three in number,
are situated about eight miles from Cairo, and should be
visited by the tourist before entering on his river cruise.
They stand on a ridge of stone, which has been so cut as
to form part of the basement.   The great Pyramid is
mainly composed of blocks of limestone brought from the

quarries on the other side of the Nile, about sixteen miles
off. It covers about 13½ acres; its present height is 456
feet; it must formerly have been about 480 feet high. Its

sides now present the appearance of irregular steps,
varying from four feet eight inches to one foot eight
inches; but it appears to have been covered originally
with a casing of polished granite; a portion of the covering still remains on the second Pyramid. Herodotus
tells us that 100,000 men were employed twenty years
in building this Pyramid, which appears to have been
chiefly intended as a mausoleum of its founder. The
granite covering on the second Pyramid makes its ascent
more dangerous than the first, which presents no other
difficulty than the ascent of a rugged staircase, about
four hundred feet in height, in which the steps vary from
two feet to a little more than four.

Near the Pyramids, more wondrous and more awful
than all else in the land of Egypt, there sits the lonely
Sphinx. This monument, so imposing in its aspect, even
in the mutilated state to which it has been reduced, has
always excited the admiration of those who possessed
sufficient knowledge of art to appreciate its merits at a
first glance. The contemplative turn of the eye, the
mild expression of the mouth, and the beautiful disposition of the drapery at the angle of the forehead sufficiently attest the admirable skill of the artist by whom
it was executed.

HELIOPOLIS.—Heliopolis, the sacred city, the On, where
Joseph's wife, Asenath, lived. A few scattered blocks, a
solitary obelisk covered with hieroglyphics, these, with
some mounds of sand and rubbish, are all that is left to
mark the site of the once priestly city.

THE SIMOOM.—In crossing the desert travelers are
frequently exposed to the Simoom or sand storm. Its
approach is indicated by a redness in the air, the sky is
suddenly overcast, clouds of hot sand obscure everything, and often render further progress for the time
impossible. The whole caravan, camels and men, then
lie prostrate on the ground till it passes over.

COLOSSAL STATUES OF THEBES.—The Colossi of the
plain. These immense sitting figures, fifty-three feet
above the plain, which has buried their pedestals, were
erected by Amunoph III, and were originally in front of
a large temple, of which only the ground-plan remains.
The more distant statue is the vocal Memnon of history.
An inscription made by one of the Roman emperors
records the hearing of musical sounds.

OBELISK AND PROPYLON LUXOR.—Part of the ruins of Thebes shows the arrangements that the Egyptians adopted in their temples. The entrance by a doorway between two immense moles of stonework, termed pylæ. The victories of Rameses are sculptured on the face of the pylon; but his colossi, solid figures of granite, which sit on either side of the entrance, have been much defaced. The lonely obelisk, seen a little in advance to the left, is more perfect than its mate, which now stands in the Place de la Concorde, at Paris.

COLOSSAL STATUE REMESES.—The mutilated statue in this view was the largest monolithic figure transported by the Egyptians from the place where it was quarried. Its weight when entire was nearly nine hundred tons, and this statue now lies in enormous fragments around its pedestal. The statue in its sitting position must have been nearly sixty feet in height, and is the largest in the world; one of its toes is a yard in length. The Turks and Arabs have cut several mill-stones out of its head without any apparent diminution of its size.

APPROACH TO THE TEMPLE AT KARNAK.—From the entrance of the temple at Luxor to the pylon at Karnak, a distance of a mile and a half, an avenue of colossal sphinxes once existed. The sphinxes have disappeared and an Arab road leads over the site. On reaching the vicinity of Karnak the camel path drops into a broad excavated avenue, lined with fragments of sphinxes. As you advance the sphinxes are better preserved and remain seated on their pedestals, but they have all been decapitated. Though of colossal proportions, they are seated so close to each other that it must have required nearly two thousand to form the double row to Luxor. The avenue finally reaches a single pylon, of majestic

proportions, built by one of the Ptolemys and covered
with profuse hieroglyphics. Passing through this, another pylon, followed by a pillared court, and a temple
built by the later Remisides.

HALL OF COLUMNS AT KARNAK.—Three thousand
years ago and this forest of columns was standing. Here
Cambyses stayed his chariot-wheels to gaze in wonder at
the triumphs of architecture. Here Sesostris was welcomed back with the loud acclaim of millions from his
conquests. The Cæsars were awed into humility when
they trod these aisles, and even the Arab hosts, as they
swept by on the tide to victory, paused to admire; and
the armies of France, as they rushed in pursuit of the
flying Memlooks, were so struck with amazement at
the ruins that they fell upon their knees in homage and
rent the air with their shouts of applause.

The main aisle is composed of an avenue of twelve
pillars, six on each side, each thirty-six feet in circumference and nearly eighty in height. Ponderous masses
of sculptured stone. The spreading bell of the lotus
blossoms crown them with an atmosphere of lightness
and grace. On each side of the main aisle are seven other
rows of columns, one hundred and twenty-two in all, of
immense size, and so close as sometimes not to allow a
column that has lost its erect position to fall to the
ground. They date from the time of Rameses III, the
Sesostris of Greek writers. These columns are a good
illustration of the way in which the Egyptians covered
all parts of their buildings with inscriptions.

THE OBELISKS AT KARNAK.—These obelisks, the most
ancient now standing in Egypt, date about 1800 B. C.
They are granite, and retain the sharpness of their
angles in a wonderful manner. This view shows in a

striking manner the desolation that prevails over all these Egyptian ruins. The total circumference of Karnak, including its numerous pylæ or gateways, is a mile and a half. The row of columns seen in the picture are part of the Hall of Columns.

THE APPROACH TO PHILÆ.—Philæ, the "Jewel of the Nile," is situated a short distance from those rapids of the Nile, known as the first cataracts. These cataracts are formed by the bed of the river being crossed by a formation of granite, through which it has cut its way, producing a series of rapids. Opposite to these cataracts stood the ancient city of Syene. It was from the quarries at Syene that the Egyptians obtained their monoliths, whether obelisks or statues. These were sculptured on the spot, and then transported by the labor of men to the places where they were to be erected. The island of Philæ contains about fifty acres, and is covered with ruins of temples and palaces, all of which belong to the Ptolemaic period. The basin of black jagged mountains folding it in on all sides, yet half disclosing the avenues to Nubia and Egypt; the clusters of palms, with here and there a pillar or wall of a temple, the ring of the bright river, no longer turbid, as in lower Egypt; of these it is the centre, as it was once the focus of their beauty.

VIEW ON THE ISLAND OF PHILÆ.—The temple which belongs to the era of the Ptolemys, and is little more than two thousand years old, was built by various monarchs, and is very irregular in its plan. The columns of the temple are very different from those of Luxor and Karnak, indicating the result of the contact of Greek and Egyptian systems of architecture. Above the true capital is a square block that bears on its four sides the head of Osiris, under the form of a bull. It was into this form of idolatry that the Israelites were so constantly lapsing, termed in Scripture, the Worship of the Golden Calf.

PHARAOH'S BED, PHILÆ.—This temple is almost perfect; it never had and never was intended to have a roof. It is one of that class termed Hypœthral temples, from their being open to the sky. Its name, Pharaoh's Bed, is derived from a tradition that Osiris was buried at Philæ, and from this it was that the Egyptians were in the habit of swearing by him who lies at Philæ.

SCULPTURED GATEWAY.—This is a good illustration of the way in which almost all parts of the buildings were covered with inscriptions. The large figures on this doorway were originally painted in bright colors, and on some of these, patches of the original paint still remain.

TEMPLE EDFOU.—This is perhaps the best specimen extant of the pylon of the Egyptian temples; it is upwards of one hundred feet in height, but a considerable part of the base is covered up with sand, which has also almost filled up the area of the temple. In this part the valley of the Nile is wider than in many places; it varies from about ten miles in width to only enough to allow of the passage of the river. Many of the temples are built close to the waters of the sacred river.

TEMPLE OF KALABSHE, NUBIA.—The space inclosed within the ruins of this temple is covered with sculptured figures, among which the most remarkable is the representation of a human sacrifice, where the victim, whose whole clothing consists of a scanty waist-cloth, is on his knees with his hands tied behind his back.

Behind him stands a priest with lofty mitre, who with one hand holds him by his long hair, while in the other he brandishes a small axe, ready to strike off his head. This horrid scene takes place in the presence of Osiris Hierax, who is seated on his throne enjoying the spectacle.

THE SHADOOF.—This view presents a scene on the Nile. A group of stately palm trees, tall and slender, with feathery plumes on their proud heads, and large clusters of golden fruit. The shadoof is a simple contrivance for raising water; a method very common both in ancient and modern Egypt. It consists of a lever moving on a pivot, which is loaded at one end with a lump of clay, or some other weight, and has at the other a bowl or basket, as seen in the picture. Wells have usually troughs of wood or stone, into which the water is emptied for the use of persons or animals coming to the well.

VIEWS OF INTEREST IN DIFFERENT PARTS OF THE WORLD.

These are described in gazetteers, and to some extent in school geographies. Some lecturers appear as very accomplished travellers by using well-written guide-books.

For an acquaintance with historical pictures, we may consult the histories of the times.

With regard to "views conveying moral lessons," the name of each slide affords a text upon which the lecturer may base what remarks he may have to offer.

NURSERY TALES.

English catalogues contain many familiar poems and stories, particularly the nursery tales, which are illustrated by lantern slides. Not having room to reprint these here, we would refer to the toy books everywhere sold for these wonderful specimens of English literature. At the risk, however, of making the rest of our matter seem prosy by contrast, we will copy just the closing part of the description of a long slide of animals, to indicate how much is made to depend on words and music, and how little on the merits of the slide.

[*Sound of Horn. Music. Last tune of the " LANCERS."*]

Yes, here we are in full cry ! The real thing, too !!

" Old Mother Slipper Slopper jumped out of bed,
  And out of the window she poked her head ;
  Husband ! O husband ! the gray goose is dead,
  And the fox is gone out of the town, O ! "

Yes, there he goes, and the old lady after him, and she has called up John, the servant, and he joins in the chase, and old Mr. Slipper Slopper comes next ; but he's rather behind, as he's been to call " Bumble," the parish constable, who has come out with his staff to catch the thief. Tally ho !

And now, my children, recollect I told you that the lion was the king of the beasts, and so, as a conclusion to this entertainment, I shall show you how he kept his court.   (*Music.*)

There he is, sitting in full state ; and now, if our kind friend at the piano will play a " March," you shall see a grand procession, and all the animals passing in order before him.

["GRAND MARCH," *during which the slide is moved slowly.*]

GOOD NIGHT.

Tune and Motto, "GOD SAVE THE QUEEN."

COMPOSITION PICTURES.

The miscellaneous views in Class X are mostly composition pictures, suggesting their own descriptions. Take, for example, this picture of the milkmaid.

The cow, so gently submitting to the maiden's manipulations, evidently feels quite at home. Appearances indicate that she is capable of giving a pailful of milk. She has taken the position convenient for the milkmaid, who, for the time, has suspended operations for a social chat with the young farmer who is resting upon the barnyard gate. We may not hear what they say, but little sister, doubtless, is verifying the old adage, that "little pitchers have large ears."

The two reclining animals may have borne the yoke seen at the left, during working hours, and are now wooing

"Tired nature's sweet restorer, balmy sleep."

The animal at the right is too young for active service, and has not yet experienced the ills of a laborious life, of which the harrow near by is a suggestive emblem.

9

The old hen in front cannot boast a very numerous brood, but the fewer mouths the better cheer.

> " Throw some crumbs and scatter seed,
> And let the hungry chickens feed."

The farmhouse on the rising ground, nestled among the trees, has an imposing appearance, but it is nothing to be compared to the elegant castles built in the air by that admiring young farmer and the loving maiden. May the course of their true love ever run smooth.

STATUARY.

Statuary and many other pictures may also be announced, and then described by what the picture itself shows, as in the example following:

THE COUNCIL OF WAR, by John Rodgers.—President Lincoln is seated and holding before him a map of the campaign. Secretary Stanton stands behind his chair, wiping his glasses and listening to General Grant, who is explaning his plan, which he is pointing out on the map.

THE SEASONS, by Thorwaldsen.—Four circular bas-reliefs, viz.:

Spring.—A female figure, attended by two genii bearing baskets of flowers.

Summer.—A harvest scene, with a group of reapers.

Autumn.—A hunter returns to his home bearing game; a woman and child (seated beneath a grape vine) receive him.

Winter.—An old man warming his hands over a brazier, while an old woman lights her lamp.

> " Behold, fond man !
> See here thy pictured life ; pass some few years,
> Thy flowering spring, thy summer's ardent strength,

Thy sober autumn fading into age,
And pale concluding winter comes at last
And shuts the scene."

MOVABLE SLIDES.

These of course tell their own story. Now and then, an appropriate recitation can be found for them.

The swan floating upon the moving waters, for instance, may be assumed as illustrating the legend that her first and only song is sung as she floats down the river on her dying day.

" 'Tis the swan, my love,
She is floating down from her native grove,
No loved one now—no nestling nigh—
She is floating down by herself to die.
Death darkens her eye and unplumes her wings,
Yet the sweetest song is the last she sings.
Live so, my love, that when Death shall come,
Swan-like and sweet, it may waft thee home."

Spectators, in the limited time given them, can hardly be expected to take in all the details of a complex view, without more or less of this particularizing, which can be resorted to as occasion requires, therefore, in connection with a wide range of subjects.

SCIENTIFIC SLIDES, &c.

The illustrations enumerated in the Scientific Department, of the appended catalogue, are suited to the text-books in common use. Works on natural history afford descriptions of beasts, birds, fishes, reptiles, and insects. Botany describes plants and flowers.

The explanations in Wells's Geology, Cutter's Physiology, &c., are just as well suited to the corresponding classes of lantern slides, because they are mostly after the same designs.

The set of long astronomical slides has from time immemorial been accompanied by a printed lecture, which, though somewhat antiquated, still answers a pretty good purpose.

Could a suitable lecture of similar shape accompany each of the forty sets of scientific illustrations, it would prove advantageous to many, and it would do no harm to any; so we are looking for something of the sort in the near future. But these sets of scientific slides themselves leave scarcely anything to be desired in the way of fitness and excellence; and we have, moreover, in the Sciopticon an instrument unrivaled for convenience combined with efficiency.

As before intimated, little has been attempted in this chapter but to indicate some of the ways of finding descriptions.

When the use of the magic lantern was very limited, its slides could be described in small compass; but now, a work that should describe all the slides in use, would hardly be less voluminous than the Encyclopedia Britannica.

CHAPTER IX.

The Sciopticon and its Uses.

DESCRIPTION OF THE SCIOPTICON.

[*From the Journal of the Franklin Institute.*]

"Our attention was drawn some time since to this very decided improvement in lanterns illuminated by ordinary flames, by which their efficiency is so greatly increased that many results can be reached which were heretofore only attainable by aid of the lime or magnesium lights.

"The most important feature in this apparatus is the lamp, or, as it might, in this case, be called, from its appearance, the furnace. This source of action to the entire machine is placed in a cylindrical chamber, provided with a chimney, and has two flat wicks, one and a half inches long, parallel to each other and to the axis of the chamber, and in fact the optical axis of the instrument. The flames, or rather sheets of flame, that rise from these wicks are drawn together by the arrangement of the draft, and so form a pointed ridge or edge of intense light in the axis of the condensers. We have, on various occasions, alluded to the fact long ago pointed out by Rumfort, that flame was practically transparent. Here this property is utilized, and by reason of it we can get through the condenser all the accumulated brightness of the long line of light, one and a half inches deep.

"We have witnessed a number of experiments with this lantern, and can fully indorse it as a great advance upon any thing before used in the shape of a lamp-illuminated magic lantern.

"For a parlor or school exhibition, it may well take the place of the far more troublesome oxy-calcium lantern, which it rivals in efficiency. -

"There are many details of construction which are of very ingenious and efficient character, among which we would specially notice the slide for pictures, by which, one picture being in use, another may be removed and exchanged, and then, by a single movement, brought into the field, while the other is in like manner ready for substitution."

THE MAGIC LANTERN FROM 1650 TO 1870.

[From the Scientific American.]

"The invention of the Magic Lantern dates back to 1650, and is attributed to Professor Kircher, a German philosopher of rare talents and extensive reputation. The instrument is simple and familiar. It is a form of the microscope. The shadows cast by the object are, by means of lenses, focused upon something capable of reflection, such as a wall or screen. No essential changes in the principles of construction have been made since the time of Kircher; but the modern improvements in lenses, lights, and pictures have raised the character of the instrument from that of a mere toy to an apparatus of the highest utility. By its employment the most wonderful forms of creation, invisible, perhaps, to the eye, are not only revealed, but reproduced in gigantic proportions, with all the marvelous truth of nature itself. The success of some of the most celebrated demonstrations of Faraday, Tyndall, Doremus, Morton, and others, was due to the skilful use of the Magic Lantern. As an educator, the employment of this instrument is rapidly extending. No school apparatus is complete without it;

and now that transparencies are so readily multiplied by photography upon glass, upon mica, or gelatin, by the printing press or the pen, it is destined to find a place in every household; for in it are combined the attractive qualities of beauty, amusement, and instruction.

" The electric light affords probably the strongest and best illumination for the Magic Lantern; then comes the magnesium light; but their use is a little troublesome and rather expensive; next to these in illuminating power is the oxy-hydrogen or Drummond light. The preparation of the gases and the use of the calcium points involve considerable skill.

"Need has long been felt for some form of the Magic Lantern having a strong light, but more easily produced than any of those just mentioned; and this has at last been accomplished, after several years' study and experiment, by Professor L. J. Marcy.

" The Sciopticon is the name of his new instrument, and from actual trial we find that it possesses many superior qualities. Its lenses are excellent, and in illuminating power its light ranks next to the oxy-hydrogen. The Sciopticon light is produced from ordinary coal oil, by an ingenious arrangement of double flames, intensifying the heat and resulting in a pencil of strong white light. Professor Marcy's instrument is the perfection of convenience, simplicity, and safety. Any one may successfully work it, and produce the most brilliant pictures upon the screen. It is peculiarly adapted for school purposes and home entertainment. Those who wish to do a good thing for young people should provide one of these instruments. Photographic transparencies of remarkable places, persons, and objects, may now be purchased at small cost, while there is no end to the variety of pictures which may be drawn by hand at

7

home, upon mica, glass, or gelatin, and then reproduced
upon the screen by the Sciopticon."

TRAVELLING BY MAGIC.

BY EDWARD L. WILSON.

Editor of the Philadelphia Photographer, and Photographic World.

Marcy's Sciopticon is what we want to give us a view
of the world at large, while seated in our own drawing-
room, enjoying all the comforts of home, and the pleasures of social intercourse.

Give us the Sciopticon, with the necessary slides,
before a screen or a white wall, and we will carry you
as fast or as slow as you wish, wherever the foot of man
has trod, in excellent and comfortable style.

First we look upon the screen and, in imagination, we
go driving along over the Union Pacific Railroad. We
visit the large cities on our way, and get as good ideas
of their grain elevators and their churches as if we stood
by their side. We see the Mormon tabernacle, and capture Brigham in person for our screen. On we go, over
the prairies, amid the buffaloes, dodging under the great
snow-sheds, climbing up the inclines of the jagged Si-
erras, and lo! (not "the poor Indian") we stand watch-
ing the gambols of the seals in San Francisco Bay,
straining our eyes to reach the summit of El Capitan in
the Yosemite Valley, listening to the rustlings of the
Bridal Veil, or clambering up the sides of "General
Grant" in the Mariposa Grove.

Or, we may glide up the Hudson, capturing the Pali-
sades, storming the Highlands, wander amid the seduc-
tive music of Trenton Falls, cross Lake George, "do"
Saratoga, "flee to the mountains," squeeze through the
Crawford Notch, clamber up Mount Willard, ascend

Mount Washington on the wonderful railway, descend
to the Glen, glide around to the Profile House, face
Eagle Cliff, kiss our hands to the "Old Man of the Mountain," shake up the echoes on the lake, and dare the
boulder in the flume, all in one half hour.

Then, after we have seen Niagara from a hundred
standpoints, views made in winter and summer, and
travelled up the Mississippi, through Watkin's Glen, in-
haled the freshness of White Sulphur Springs, wandered
among the wildernesses of North Carolina, and seen Flor-
ida and Cuba, not to forget the Mammoth Cave, we may
go over to Europe. There we ascend the Alps with Prof.
Tyndall, go down into the caverns, and clamber among
the icicles, or traverse the awful glaciers with their
yawning, ever-hungry crevices.

Or we may see in the same way the ruins of India,
the mysteries of Pompeii, the tombs and pyramids of
Egypt, or Rome's seven hills covered with glories, to
say nothing of humiliated Paris or exultant Germany.

Everything that photography can produce may be
served up in excellent style, and with little trouble
through the instrumentality of Marcy's Improved Magic
Lantern. Last evening I had the pleasure of entertain-
ing and delighting a whole company of men, women,
and children for an hour or two in this way, at the extreme cost of five cents for coal oil!

The great efficiency of the Sciopticon, as compared
with any other lamp-illuminated lantern, together with
its simplicity, symmetry, and compactness, its safety,
convenience, and fitness for slides of every variety and
for various philosophical experiments, makes it unri-
valled for home and school purposes.

No doubt Mr. Marcy's explanation of it will be appreciated, and I need only add that I would not want

to be without a Sciopticon in my house. It gives one such enlarged views of everything.

SCIOPTICON FOR SUNDAY-SCHOOLS.

BY E. D. JONES, ESQ.,

President Missouri State Sunday-School Association.

"While the great aim of all Sunday-school effort is to teach the word of God, seek the conversion of scholars, and train such in the ways of holy living, yet there are appliances and helps that may be used to attract and interest young minds where they do not in any way conflict with the grand object of the school.

"It is a religious institution, and its interests should be well guarded from all that would in any way lower the dignity of its mission. Some time since I introduced the Sciopticon, a recent improvement in the line of the Magic Lanterns, of which Prof. L. J. Marcy, of Philadelphia, is the patentee.

"I found the instrument wonderfully simple in construction and management. Its lamp burns simple coal oil and gives a most intense light, and in the production of pictures on the wall or on the screen equals any of the most expensive Magic Lanterns, with calcium lights, that cost so much labor and expense."

SCIENCE AT HOME.

(Communication from the President of Franklin Institute.)

Mr. L. J. MARCY.

DEAR SIR: During the winter of 1872-73 1 was interested in lantern experiments, using the lime light as the source of illumination. At the same time I made frequent use of your very admirable Sciopticon, with oil lamps. The readiness with which it can be adjusted and made ready for use impressed me. For parlor use, as a magic lantern, I very much preferred it on this account, to the more troublesome lime light. Its convenience recommends it as an adjunct to the school-room and I found that very many of the most interesting experiments in physics, usually shown in a lantern, can be readily performed with the Sciopticon. My good friend, Prof. Henry Morton, of the Stevens Institute of Technology, in Hoboken, has already described many of these experiments in your manual. I have told you how I have repeated many of them with very little expense in the way of apparatus, and I would now suggest to the would-be purchasers of your lanterns, that should they desire to use it as an adjunct to the lecture table, they need not be alarmed at the expenditure needed to procure all the fixtures required to perfect it. One of the chief pleasures in its use is in the improvising of what is needed. Those who have long purses may prefer to purchase all needed pieces of apparatus, ready-made to their hand, but a few hints may serve to show how they can, with very little skill, prepare what will answer their purpose. As an illustration, let me recall the very pretty experiment usually called the broken arrow, which is shown to illustrate refraction. As an object in the lantern, a brass plate having an arrow-shaped opening in it (procurable at the instrument makers) is put in place, this throws upon the screen a white arrow on a dark ground; now, if in front of the brass plate a strip of thick glass, narrower than the length of the arrow, be held parallel with its surface, no distortion of the arrow image will be seen; but if the glass be inclined so that the rays of light pass through it obliquely, a piece of the arrow will seem to be cut out and be moved to one side. This is a striking illustration and can be improvised quite readily, as follows: Procure some slips of

good window glass, of the size used for magic lantern
slides (I prefer 3 x 4), some tin-foil, such as paper-hangers
paste on damp walls before papering, and some paste
made of gum tragacanth; with a sharp knife, laying
the foil on a plate of glass, the arrow-shaped opening
can be readily cut, and its edges will be as smooth as
the most skilful mechanic can make a brass plate. This
foil, so prepared, should be mounted between two slips
of glass, and the edges bound with paper. Gum tragacanth will cause paper to adhere to glass very firmly and
is a nice, clean paste to use. The slide thus prepared
will be found to be quite as good as the most costly one
procurable in the stores. In my own experiments, when
I require slits or openings of any required shape, in
opaque plates, I have invariably made them in this
manner, with a feeling of satisfaction at their cheapness.

A very convenient device to show wave motion can
be made with this tin foil. One slide is made with plates
of glass, 3 x 4 inches, having tin foil inclosed, in which slits
are cut crossways, say $\frac{1}{18}$ inch wide, 2 inches long, and
the slits placed $\frac{1}{8}$ of an inch apart. I have sometimes
pasted slips of tin foil $\frac{1}{8}$ of an inch across the plate, at
equal distances, say $\frac{1}{18}$ of an inch, in preference to cutting
them in a solid piece of foil. This slide will show vertical bars of light on the screen. If now another slide be
made of two glasses, 3 x 6 inches, with foil between them,
in which foil a wave-like opening be cut, say $\frac{1}{8}$ of an inch
wide, this slide of itself would show in the lantern a wave
line of white on a dark ground on the screen. The two
slides put together in the lantern will show a wave line
of dots, and if the wave-line slide, which is twice as long
as the one with bars, be moved back and forth in front
of the bars, the dots will seem to rise and fall in wave
motions, and the fact will be demonstrated, that in wave

motions there is an advancement of the wave, while the individual particles only rise and fall without advancing.

The slips of glass, mentioned above, can be conveniently prepared for drawing diagrams, by coating one side with plain collodion (gun cotton dissolved in equal parts of alcohol and ether); when dry this surface takes India-ink admirably, and diagrams can be traced, or pictures copied in a rough way, by laying the glass plate so prepared over the picture to be copied and tracing its outline with a pen filled with good India-ink.

I would strongly advise any one using your lantern to procure some of the comic slides, such as you illustrate in Class XV of your catalogue of slides, and they can see how to make similar ones to be used in illustrations of scientific subjects. Thus with the wreck of one of these three glass slides, picked up at some opticians and purchased for a few cents, I improvised a slide which answered better to illustrate the process of carbon printing in photography than the process itself would have done in a lecture-room. One figure changed with another by means of sliding glass plates is very useful in many kinds of experiments or illustrations of facts and processes.

The tank figured in your manual, in Chapter VII, on Chemical Experiments, contributed by Prof. Morton, can be made to do service in a long line of experiments with electricity, by a very simple device. Thus, to illustrate the decomposition of water, cut a slip of segar-box wood, of a size that will lay on the bottom of the tank loosely, attach to this bit of wood copper wires, which will extend up to the end of the tank and will not quite meet at the centre of the bit of wood; to upturned ends at this place, solder little slips of platina foil, $\frac{3}{4}$ inch long by $\frac{1}{4}$ inch wide, they must stand vertically face to face,

about ⅓ inch apart. Now coat the copper wires and the
wood with melted paraffine, but take care that none gets
on the platina; this will insulate the copper wires and
prevent the wood from absorbing any moisture. This
little frame placed in the tank, immersed in acidulated
water (water with a few drops of sulphuric acid), and
the terminal wires attached to say two cells of Groves'
battery, will show the decomposition of water admirably.
A similar piece of apparatus with the terminal wires at
the centre of the board, united by a vertical coil of very
fine platina wire, will be found useful in illustrations of
circulation by heat. Such a frame immersed in clear
water will be seen on the screen as a black coil, seem-
ingly hanging down from a black bar on the top of the
screen; if now, by means of a pipette, some colored
fluid, say a solution of permanganate of potash in water,
be carried to the bottom of the tank, it will on the screen
seem to spread itself out as a red stripe under the black
one and enveloping the little coil; a current of electricity passed through the wire will heat the little platina
coil and thus heat the water in contact with it, so that
currents will be established in the fluid, carrying with
them the colored fluid in a very beautiful curling cloud
of color.

I mention to you these few examples, of how readily
the needful appliances for illustrations can be improvised;
now I have frequently heard persons say that they
"feared the expense entailed in the use of a lantern·"
that "the lantern is so useless without a great many
accessories, which are so expensive in themselves."
Feeling a lively interest in your very meritorious invention, I have volunteered these hints, which I beg you
will, if you see fit, use to your benefit.

Very truly yours,
COLEMAN SELLERS.

CHAPTER X.

The Teacher his own Artist.

A PROCESS FOR DRAWING AND PAINTING MAGIC LANTERN SLIDES.

The following process is given to assist persons who own a Sciopticon, to prepare for themselves a portion, at least, of the necessary transparencies, especially educational illustrations. It has been our aim to simplify the whole matter as much as possible consistently with giving such directions as are safe and practicable. It is true that there are some other colors and materials which can be used, but the list given below contains *all* that are *necessary* for the production of hand-made pictures, or for coloring photographs on glass in this style.

LIST OF APPARATUS AND MATERIALS.

Easel,
Glass slab,
Palette-knife,
Sable pencils,
Duster,
Point for erasing,
Hard black-lead pencil,
Fine pen,
Varnish, Nos. 1 and 2,
Liq. India ink,
Canada balsam,

Siccatif,
Tube of blue-black in oil,
Tube of crimson lake in oil,
Tube of Italian pink in oil,
Tube of Prussian blue in oil,
Tube of burnt sienna in oil,
Ol. Turpentine,
Glass,
Mats,
Binding-paper,
Box for the above articles.

For amateur work we very decidedly recommend that the pictures should be made on the 1-4 size of glass known to photographic stockdealers as " B. P. C."

After the painting is finished and dry, it is only necessary to cover with a mat and another piece of the same glass, binding the edges with narrow strips of paper, to give us the form of the " Woodbury," or of the new " Crystal " slide.

If the glass which we recommend is used, simply

breathing on and thoroughly rubbing it with tissue-paper is sufficient for the cleaning. Of course the fingers must not touch the surface of the glass after it is cleaned.

The next step is to prepare the surface of the glass for the drawing. If the glass (as is usual with this kind) is slightly curved, the *concave* side should invariably be the surface to receive the preparation. This preparation is as follows : "The plate to be dusted and gently warmed ; then flow the surface with No. 1 varnish, and drain into the bottle from the corner of the glass. When this is dry, flow with No. 2 in the same manner, and afterwards dry with gentle heat. Of course until the varnish is dry the corner from which the varnish was drained should be kept down. Should dust find its way into the varnish, it can be filtered." Having prepared the surface, it should not be soiled by handling (always take the glass by the edge between the fingers). A sketch can now be made on the surface with a good hard *black*-lead pencil, either as an original drawing, with the glass on the easel, or by placing the glass on an engraving or other picture, and tracing the outline.

When this sketch or outline is finished, strong and black lines and marks can be made with the "liquid India-ink" in a fine pen, gently used so as not to make the lines too coarse and heavy. With the pen, too, lines can be ruled or letters or figures added. After the drawing is finished and dry, any little corrections or scratches can be made with the pointed eraser.

The easel should be so placed that the light falling on the white tablet will be reflected to the eye of the artist through the inclined glass on which the picture rests, and consequently the picture will be illuminated by *transmitted* light.

The easel may stand on a table placed against a window into which the sun does not directly shine. The little screen which fits the easel is to guard the eyes of the artist from extra light which comes from above the easel.

Although the pigments in our list are so few in number, yet with these, by proper admixture, *all* the colors can be made.

To mix these colors for painting we will suppose we take from one of the tubes a quantity of color of the size of a small pea, to which we add about half as much Canada balsam, and one or two drops of siccatif. The use of the balsam is to make the colors *transparent;* the siccatif is simply a *drier,* and of this last we should always use as little as is consistent with the colors drying in a reasonable time.

If a picture involving a variety of colors is to be painted, it is best before commencing the work to prepare not only the simple colors in the tubes, but also some of the most important mixtures, as follows:

Greens.—Prussian blue and Italian pink. (To make a dull green, use some burnt sienna with the above.)

Orange and Flesh-tints.—Italian pink and crimson lake. For some shades add burnt sienna.

Browns.—Italian pink, Prussian blue, crimson lake, and burnt sienna.

Any shade of brown or neutral tint can be made by the use of these pigments in various proportions, which can only be learned by experiment.

Having prepared the colors on the glass slab (mixing well with the palette-knife), and the drawing being all ready, after dusting we proceed to apply the first coloring.

If the subject is a landscape, the first thing to paint is the sky. A little color, slightly thinned with ol. turpentine, can be applied with two or three strokes of a wide brush. It does not answer very well to torture the color with the brush after it is once applied, so we must be able to do what we wish with a *very* few bold strokes. We don't mind about trees, spires, &c., we can take the sky tint off from them where we wish to do so afterwards (before the color is dry) with a sable brush, *slightly* moistened with ol. turpentine, but the brush must not contain enough to *spread at all* on the surface, otherwise the color will flow away from the line we wish to establish, and form a ridge on the sky.

A better way to apply sky tints is to stipple or daub the surface with a little paint on the end of the finger. In this way all the finest skies are painted by the best artists who make pictures for the magic lantern, but the process is one that can only be learned by many trials and much practice.

After the sky is painted and removed from places where it is not wanted, distant hills next receive our attention, successively working upon objects nearer and nearer, until the whole picture has received its first painting. Except for skies the colors will not need thinning with ol. turpentine.

The picture should now be put out of the way of dust and sunshine, and where it will dry. The next day such parts as need can be repainted, and, if necessary, any successive number of coats can be applied, allowing a day between each for the paint to dry. When completed, it only remains to put on a mat and cover and fasten the edges with paper strips. *Flour paste* (too stiff to allow moisture to be drawn up between the plates) is the best.

When in use, that is when changing from one color to another, the brushes can be cleaned with ol. turpentine. But when the brushes are to be put away, they should be washed with warm water and soap, rinsed with *clean* water, and then brought to a point to dry.

The preceding article on drawing and painting magic lantern slides has been contributed by an expert preeminently qualified to assist us in overcoming its apparently insuperable difficulties.

It may be proper here to state that we now furnish a varnish to take the place of No. 1 and No. 2, which answers equally well with but once flowing. Like No. 2 it must be dried by heat to prevent what is called "chilling." The operation can be best performed in a dry atmosphere which is free from dust.

A COPYING CAMERA.—A private letter from Prof. W. A. Boles, of Shelbyville, Ind., Superintendent of Schools, contains further valuable information in this direction, and a description of a new instrument of his own invention. By permission the following extracts are given in the interest of home production.

"With this mail I send you a specimen of my drawing on gelatine-coated glass, for use in the Sciopticon. After the coating of gelatine is perfectly dry, I sand-paper it with the finest article I can get, and after the picture is drawn, float it with your No. 1 varnish. . . .

"I made an upright camera-obscura, using the lens from the Sciopticon. The picture to be copied is placed beneath, in the sunlight if possible, and the image is thrown upward so that I can trace the outlines quite at my ease. On the roughened gelatine a fine steel pen and the ordinary black ink I am now using mark beautifully, and the shading is done with a lead pencil. By tracing the image of the picture, in the little darkened chamber, I avoid the trouble from the slipping of the glass and the different angles of observation consequent upon superposition. This plan has also the additional advantage of enabling me to reduce or enlarge a drawing to any desirable size.

"The accompanying rude drawing will give you some idea of the plan.

Fig. 31.

"The front A and the rest B are connected by two iron braces and slide up and down, for the desired focal distance, above the lens C. The adjustable shelf D supports the object to be copied. The glass plate is laid upon the rest B over the opening, and a black cloth thrown over the top. The instrument is six feet in height, and eighteen inches square, and cost me about two days' labor. It is highly satisfactory."

It will be understood that the object of roughing the gelatine, as indicated in the above letter, is that it may present an opaque surface to render the image visible, like the focusing ground-glass in a photographer's camera; this roughing also gives it a "tooth" to hold the markings.

Ground-glass itself is sometimes used for obtaining drawings by superposition, which is rendered transparent by a coat of varnish, and it might be used in this case; it is, however, somewhat expensive. A surface is now produced on glass quite similar by what is called the ground-glass varnish. This surface after receiving the drawing may also be rendered transparent by a coat of common varnish, if dried in sufficient heat to prevent its again becoming opaque.

ASPHALTUM VARNISH.—This common black varnish, flowed upon quarter-plate glass and allowed to dry, gives a surface which can easily be drawn upon with a sharp instrument so as to show upon the screen in smooth white lines upon a dark ground. The varnish must be thinned with turpentine, to allow of tracing by superposition.

This process, obtained by purchase, is given for the benefit of teachers, who will find it admirably suited for maps, diagrams, and any so-called blackboard exercises.

ORDINARY TRANSFERS UPON GLASS.—Coat, by preference, quarter-plate B. P. C. glass on the hollow side with transfer varnish, and let it partially dry. Take any cut of proper size, place it upon water right side up until it becomes wet to saturation, adjust it to the varnished surface face down, rub up the paper in rolls until it is mostly removed, and then lay it aside until the varnish becomes hard. Much of the remaining paper can now be removed without damaging the picture, by carefully wetting and rubbing.

Grecian varnish will render the picture transparent; but unfortunately innumerable pimples make their appearance, which show badly on the screen, caused by the loosened fibres of the remaining paper film. Several coats of varnish will finally leave a smooth surface, but so thick a coating will before long crack and eventually peel off.

This process is here referred to because it is so often inquired about, rather than because it is thought to be of much value.

DIAPHANIE.—A picture on thin, smooth paper, treated as above, but without rubbing off any of the back surface or even roughing it up, will make a tolerable lantern slide, with one coat of the varnish. It will appear dark on the screen like a silver print that is very much "fogged." If the picture is pasted to a frame without glass, and varnished, it appears rather better with less cost.

DECALCOMANIA, OR ENAMEL SLIDES.—Impressions made on starched paper and fixed to glass, as above described, will adhere to the varnish when the paper is afterwards wet and pulled off entire. After soaking off the starch, by flowing water and a soft brush, the picture is dried and flowed with ordinary slide varnish. These so-called Enamel Slides are inferior to silver prints, and are sold, both plain and colored, at a cheaper rate.

Should the paper prints, in good variety, eventually be sold to the public, full directions for transferring them would doubtless also be supplied. On this supposition only, would this truly interesting process promise to become available for amateurs.

TYPE PRINTING upon glass, except in a poor way with elastic type, is impracticable. Hymns, mottoes, &c., may, however, be printed to advantage upon sheet gelatine, in the small amateur printing-presses now so much in vogue. Plain collodion films, dried upon oiled glass, upon a levelling stand, and peeled off, will take impressions perfectly. These films may be mounted between glass plates, in the form of a crystal slide.

WITH QUARTER-PLATE GLASS prepared as described to receive, like paper, not only colors but pen and pencil drawings, we may copy engravings for the lantern by superposition, or in a camera similar to the one illustrated by Fig. 31, may show up, as by magic, all that class of illustrations now in vogue as "blackboard exercises;" may suit original designs to the various circumstances of time and place; may transcribe hymns, mottoes, &c., appropriate to each occasion; and so, with small expense and ordinary skill, the Sciopticon becomes readily available, without the purchase of a large assortment of slides.

Fig. 32.

The cut (Fig. 32) shows the exact size and shape of the transparency, mat, glass, and the binding of the Woodbury (photo-relief) magic lantern slide, of the colored crystal slide, and of such as are made according to the directions given in this chapter.

Pen and pencil drawings would require only Nos. 7, 8, 9, 10, 19, 20, and 21, of the materials enumerated on page 138; or, simpler still, the glass may be obtained, ready prepared, for 75 cents a dozen, the drawings made with any quite black ink and a fine pen, and shown without glass cover, mat, or binding.

CHAPTER XI.

NOVELTIES.

THE CHIMNEY CAP of the Sciopticon now telescopes
into its base, so as to give added length and greater
draft when drawn up. It should not be elongated, however, when the instrument is first lighted—especially if
the wicks are not well saturated with oil—but when
well under way, the added length, with a corresponding turning up of the wicks, gives greater brightness,
and more effectually draws off the heat.

THE SCIOPTICON CURTAIN.—Turning the milled head
at either side, gives the appearance upon the screen of
a curtain rising, or falling, thus handsomely opening
or closing an exhibition. It may also be temporarily
closed at any time, to allow the attention to be directed
to other exercises.

The process of changing the pictures may be hidden
from view by shutting off the light with the left hand;
then pushing the out-going picture into the left hand
by sliding another into its place with the right;· and
then flashing on the light with the right hand; all of
which may be sooner done than said. In any change of
programme the awkwardness of showing the "full moon,"
or the disk without a picture, may always be avoided by
using the opaque curtain.

This curtain also serves as a back cap for the objective, protecting the back lens from dust and light when
not in use, as the front cap protects the front lens.

THE SCIOPTICON TINTERS.—The tinting-glasses are drawn up close behind the objective lens by means of rods terminating in knobs above. This lets the color down upon the screen—not with a sharp outline like the curtain, but with a gradual shading. With the blue tint partially drawn, this property gives to plain photographs of scenery, a blue sky, shading off without abruptness down to the horizon. Slightly drawing up the blue, then the red, and then turning the button attached to the opaque curtain a little, fades away gradually the upper portion of the disk, as is desirable in such slides as the Ascension. The reverse movements bring into view gradually the "Soldier's Dream," "Angel of Peace," &c., nearly as well as with two slides in the dissolving lanterns. All the appearances formerly produced by colored glass slides are better effected by these tinters. If at any time the rods become too loose, the *stop* screws may be tightened.

THE NEW SLIDE STOP.—The catch drawn out from the underside of the stage, and turned back almost out of the way, is intended to stop the slide in its proper position, but not to interfere with its being moved smoothly along and out by the incoming slide when slipped forward from the catch by the left hand.

The catch pushed back into its sheath, leaves the stage entirely unobstructed for those who prefer to adjust the slides by hand only, and for other than wooden slides.

The use of stops is the more necessary with a pair of instruments, as it is essential to a proper effect that the dissolving views should occupy exactly the same place on the screen without any readjustment after the dissolving becomes visible.

LARGE SLIDES— A NEW DEPARTURE.—To improve the size and brilliancy of a Sciopticon exhibition, the expedient, next to perfecting the instrument itself, is to bring into use larger and clearer views.

A magic lantern picture of the standard size is 3 inches in diameter, mounted in a frame 7 inches long by 4 inches wide. The new picture is 3½ inches in the clear, in a frame 7 inches by 4½ inches.

The new picture having a third more surface, the illuminated disk shows larger in proportion and to very much better advantage. No one seeing an exhibition of the new slides, would willingly select from the old.

No inconvenience arises from using both kinds in the same exhibition. The larger frame, reaching a little higher than where the spring meets the condenser, slides more smoothly into place.

Great pains has been taken to get the best subjects with which to inaugurate this new departure, and to have them worked up in the most artistic style.

MARCY'S EIDOTROPE.—Two disks of perforated tin are mounted so that one extends beyond the frame to the right, and the other to the left; and so, not having a common centre, an eccentric revolution is given to each, little by little, when moved by the fingers at either side of the lantern, producing upon the screen a great variety of strikingly beautiful patterns, which may be pleasingly modified by a varying use of the tinters. Its simplicity, however, may be against it, for *effects* are quite apt to be valued in proportion to their cost and trouble.

WHEEL OF LIFE.—In the English Wheel of Life, in its best form, the opaque disk with the open section, as seen in Fig. 33, is revolved rapidly, while the transparent disk, covered with figures in varied positions, is revolved with less rapidity in the opposite direction.

Our improvement consists in having the distance of the crank wheel adjustable, and in having only one band doubled back to run both wheels, so that the tension of the band can be regulated, while both effect wheels are subject to a band of like tension.

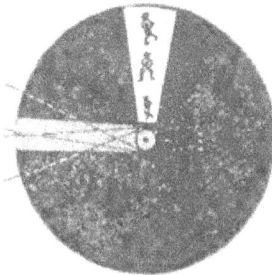

Four adjustable disks of figures, viz.: the Fishers, the Skaters, the

Fig. 33.

Giant's Ladder, and the Bottle Imp, are now included in this new apparatus; each, when used, being held in place by a wire ring sprung into a surrounding groove.

THE DANCING SKELETON.—This effect is produced by having, say six transparencies of a skeleton, in as many different postures, set in a large disk, as shown in Fig. 34. Giving this wheel ⅙ of a revolution at a time, brings the figures one by one into position to be projected upon the screen, while a revolving opaque disk hides the passing off of one, and the coming on of another, making it appear as if the same figure

Fig. 34.

were keeping time to the accompanying music, by striking grotesque attitudes.

The thousands who witnessed this *striking* feature of Prof. Pepper's late American lecture tour, seemed fully to enjoy the worth of their money.

It is but fair, however, to observe that the machine is complicated and cumbersome, and that it doubtless fascinates more for its novelty than for its intrinsic value

Having the only dancing skeleton in America, and that well enveloped in mystery, is quite different from having the apparatus explained in print, and offered for sale at $30.

THE DANCE OF THE WITCHES.—Paper witches are put into the cell (Fig. 35) and the picture of a cauldron is drawn upon its outer face. This cell is filled with water, which, with the floating witches, is made to circulate about the pot by means of pipettes with rubber bulbs, reminding us of the scene in Macbeth :—

Fig. 35.

" Roundabout the cauldron go ;
  In the poisoned entrails throw,—
  Double, double, toil and trouble,
  Fire burn, and cauldron bubble."

THE BEEHIVE.—Upon the fixed glass, a beehive and shrubbery are artistically drawn and colored. Two disks of glass covered thickly with figures of bees on the wing are revolved in opposite directions, presenting upon the screen a very lively appearance.

THE AQUARIUM.—Upon the fixed glass, an aquarium with dark background is drawn, while fish are made to appear as swimming back and forth, passing and repassing each other, by revolving in opposite directions, the disks upon which they are painted.

THE FOUNTAIN.—The appearance of a fountain in full play is produced by revolving over the face of the picture disks of glass covered with crinkles.

THE ROCK OF AGES.—A dark and stormy sky, and the waves dashing against a stone cross (the Rock of Ages), are here represented. A wreck is seen in the distance. The wreck disappears, and the figure of a woman appears clinging to the cross.

The following four pairs of effect slides for the double lantern, with sliding movements (price $6 a pair), are quite popular at present, but they are difficult to operate satisfactorily.

THE NAIAD QUEEN.—View of a lake by moonlight. The moon glistens on the water; a castle is seen in the distance. The Naiad Queen appears sailing across the lake in a pearl shell boat and playing on a harp. (Music can be used very effectively with this view.) Two slides, with movement.

TRAIN OF CARS.—A railroad bridge in a dark forest is seen by moonlight. A train of cars dashes by, the headlight, and sparks flying from the engine, making a very brilliant appearance. Two slides, with movement.

OCEAN STEAMER.—A view of a harbor is represented, with a city in the distance, from which an ocean steamship, bound for Europe, sails away. Two slides, with movement.

THE SERENADE.—A beautiful view by moonlight of a lake, on the borders of which is seen a castle brilliantly illuminated. The serenader appears sailing in a gondola and playing a guitar. A lady steps out upon the balcony of the castle and listens to the serenade. (Music can be used in connection with this effect.) Two slides, with movement.

Each of the following effects is shown either with one slide in a single lantern, by means of tinters and curtains, or with two slides in a pair of dissolving lanterns.

WASHINGTON'S DREAM.—Washington, tired by the duties of the day, in his room seated by the table, has fallen asleep. The camp is seen through the open door. The vision of Columbia, attended by Justice and Plenty, appears in the sky.

THE SOLDIER'S DREAM.—This is best told in Campbell's Poem, beginning:

> "Our bugles sang truce, for the night cloud had lowered,
>   And the sentinel stars set their watch in the sky;
> And thousands had sunk on the ground overpowered,
>   The weary to sleep and the wounded to die.
> When reposing that night on my pallet of straw,
>   By the wolf-scaring faggot that guarded the slain,
> At the dead of the night a sweet vision I saw,
>   And thrice ere the morning I dreamt it again."

MERCY'S DREAM.—Mercy is represented in a reclining position beneath a spreading tree. An angel from Heaven appears and places a crown of glory on her head.

ANGEL OF PEACE.—A beautiful landscape showing a city at night, with the new moon in the sky reflected in the water. The figure of an angel bearing a child appears like a vision in the sky, and then fades away.

THE FAIRY GROTTO.—A view looking out from a grotto into a lake, upon the surface of which are interspersed rocks and trees in a picturesque manner. A fairy with wand is seen in the foreground. The moon appears from behind the clouds and its reflection is seen on the rippling waters.

THE WIZARD'S GLEN.—The wizard and his attendant fairy are seen in the foreground of a picturesque glen in the mountains. The moon glistens on the crest of a cascade, which falls from the summit of the rocks.

CHAPTER XII.

The Lime Light. •

INTRODUCTION.

THE Lime Light in an improved form having been introduced into the Sciopticon, it becomes expedient to append to the Sciopticon Manual a description of the apparatus and directions for its use.

FLAME ILLUMINATION.

We have in the Sciopticon oil lamp arrangement the largest amount of ordinary flame illumination that can, to advantage, be brought to bear upon the screen through the objective lens, and as bright as air with its twenty per cent. of oxygen will make it.

Brightening the two flames by an inner supply of oxygen gas, after the manner of the Bude light, heats the wick tubes to an unsafe degree, and consumes too much oxygen as compared with the efficiency of the illumination.

Some solid matter, of which quick-lime seems to be the best and cheapest for the purpose, is more luminous in an intense heat, than simply the particles of burning carbon floating off in flame.

VARIETIES IN LIME LIGHT.

Every lime light alike has lime and a jet of oxygen gas; but a variety arises from the different conditions of the hydrogen employed. With alcohol vapor, it is commonly called the oxy-calcium; with house gas led directly from the bracket, it is sometimes called oxy-hydro-calcium. To produce the light more generally known as the oxy-hydrogen, both gases are forced upon the lime from reservoirs under equal pressure, and through a common jet.

In absence of well established and distinctive terms, we seem obliged to fall back upon the only elementary distinctions, as alcohol, house gas, and mixed jet; or simply Nos. 1, 2, and 3.

THE ALCOHOL BURNER.

The lime light produced by a jet of oxygen gas through an alcohol flame, as now used with improved effect in the Sciopticon, is suited, next to the Sciopticon oil lamp, to the widest range of circumstances.

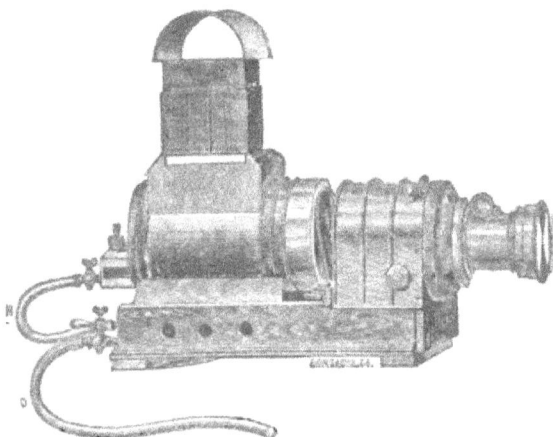

Fig. 36.

The alcohol fountain at the side of the Sciopticon is shown in the above cut as drawn back, so as to be partly seen. It is connected with the triple blow-pipe at the hydrogen stop-cock by a piece of rubber tubing, *H*. The tubing, in connection with the oxygen stop-cock, extends downward towards its connection with the oxygen bag, or cylinder, which is not here shown.

The alcohol passes through a side aperture, *h*, up the nozzle *n* (Fig. 37), to its level in the cylindrical wick-tube *a*, filling it about two-thirds full, and saturating the wick *w*, which loosely fills the space between the concentric tubes.

If on lighting the wick (which is done by reaching it in the lantern with a lighted match) it burns feebly, for want of being fully saturated, we may give the flexible alcohol tubing, *H* (Fig. 36), a sliding pressure towards the jet, being careful not to overflow the wick tube. The alcohol should be of the best quality, and high enough in the fountain to feed a vigorous flame.

The atmosphere, while supplying oxygen to the outside of the flame, compresses the inner hydrogen vapor into combination with the jet of oxygen as it strikes upon the lime above.

No. 1, Fig. 37.

When the exhibition is over, we may let the fountain hang down by its tubing till the alcohol drains back into it, before closing the stop-cock.

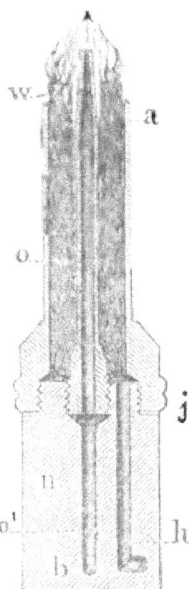

THE HOUSE GAS BURNER.

In towns and wherever illuminating gas is in supply, it may be led by flexible tubing from the bracket to the blow-pipe, and so through the same side aperture, *h*, in the nozzle into the conical tube *g* (Fig. 38), burning above the oxygen jet *o* like the alcohol flame in No. 1.

11

While the amount of the alcohol flame depends on the height of the fountain, *this flame* is easily regulated by the hydrogen key to the exact amount required by the oxygen.

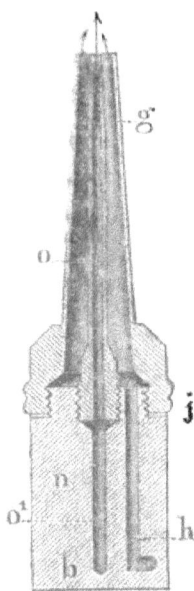

The oxygen, shooting up like the middle arrow from the *flat-mouthed* jet, *o* (see its shape, front view, Fig. 39), combines with the hydrogen instantaneously as it strikes the lime, with little tendency to cool and blacken a central spot.

THE DRUMMOND LIGHT.

Lime rendered incandescent by the ignited jet of an oxy-hydrogen blowpipe, invented by Dr. Hare, of Philadelphia, and used with marked success in the British Signal Service, by Lieut-Drummond, has been called the Drummond Light.

No. 2, Fig. **38.**

This term applies distinctively when a collected supply of hydrogen gas, as well as of oxygen, is required, and when both are alike under heavy pressure and forced through the tubes of the blow-pipe.

Oxygen and hydrogen, mixed in a common reservoir, cannot be safely used in connection with an ignited jet.

The separate gases, however, may be forced upon the lime through entirely separate jets, in accordance with the primitive arrangement; or, which amounts to about the same in effect, the two currents may terminate in tubes one within the other, called the concentric jet, mingling as before only when projected upon the lime; or, which is counted the most effective and is now in general use, the oxygen and hydrogen may be mixed in a small chamber, as at *c* (Fig. 39), at the foot of a single jet through which the mingled gas is projected upon the lime.

THE MIXED JET.

The mixed jet is the only one of the Drummond variety superior in effect to Nos. 1 and 2, and this we adopt as the No. 3 of the triple blow-pipe.

The mixed jet tube, *m* (Fig. 39), takes the place of both the tubes of No. 2, covering both apertures at the junction, *j*.

The oxygen gas forced up through the central aperture *o'*, in the nozzle, instead of being taken direct to the lime by the jet tube, as in Nos. 1 and 2, is allowed first to mingle with the hydrogen forced in with equal pressure at the side aperture.

The relative quantity of each gas is regulated by the stop-cocks till the best effect is produced, which, in theory, is when there are two volumes of pure hydrogen to one of oxygen, or about equal volumes if coal gas is used.

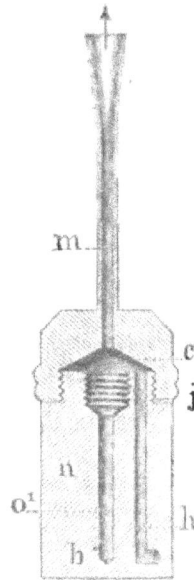

No. 3, Fig. 39.

DANGERS PECULIAR TO No. 3.

Serious accidents in operating the mixed jet are mostly occasioned by carelessly collecting or transferring one kind of gas into a bag partly filled with another kind, and then attempting to use it in connection with an ignited jet. It is common to distinguish the bags by the letters *H* and *O*, and it is further recommended to use a characteristic stop-cock on each, so that there can be no mistaking them, even in the dark.

By very unequal pressure, gas from one bag may be forced through the cavity, at the foot of a clogged jet, into the other bag, from which it meets no counter current. To get up an explosion in this way would require patience, and either a surprising degree of carelessness, or else considerable skill directed to this end.

It should become habitual to turn off the oxygen at least, when the light goes out, and to turn it on only after the hydrogen is lighted.

Stuffing the cavity of a mixed jet with wire gauze is now discarded as not only useless but as often interfering with its successful working.

Passing the gas through a small wash bottle is designed to prevent the back flow of gas or flame. The same advantage is claimed for a valve in the tubing, open towards the jet but closing against any back current.

The best expedient, however, is to use good apparatus, and to exercise common care.   •

ANNOYANCES PECULIAR TO **THE** MIXED **JET.**

The hydrogen flame is somewhat liable to be blown out by too abruptly turning on the oxygen. In this case we have but to turn off the gas and proceed again with more care.

The mixed gas in the cavity c sometimes explodes and the ignition may continue within the cavity. Whether the flame is extinguished by the explosion or not, the oxygen should be at once shut off; the jet, if hot, should be cooled, by waiting or by wetting it, and the adjustments should be revised and regulated. This accident may result from irregular pressure, from turning on the oxygen before lighting, or from neglecting to turn it off before re-adjusting the pressure boards.

None of the accidents or annoyances above enumerated have ever happened with the jet herein described.
Its characteristic flat mouth, small cavity, and ready
adjustments are in its favor, but may not be relied on
in spite of bad conditions. It will be observed that
this possibility of evil in the mixed jet, as distinguished
from Nos. 1 and 2, arises from the necessity of having
two gases in bulk, and from mixing them before coming
to the outlet.

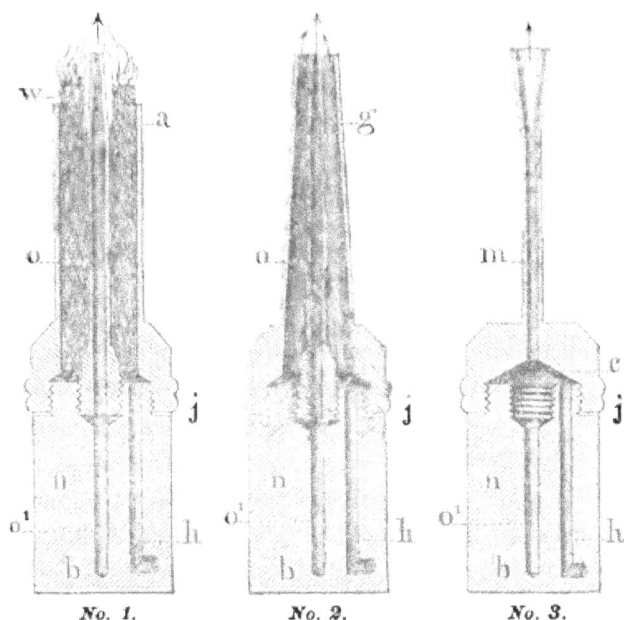

No. 1.   No. 2.   No. 3.

THE THREE JETS presented above, side by side, show
what is peculiar to each. The oxygen tube of each is
screwed down upon its lead washer, so as to present its
flat mouth to the lime, as shown in No. 3.

When a very strong current meets with roughness at
the mouth of a jet, or strikes some impediment, or an

unsound place in the lime, it sometimes produces a his-
sing sound.   Moving the lime or varying the pressure
will, in most cases, abate the annoyance.

MARCY'S TRIPLE JET.

In this apparatus, the three burners already described
are interchangeable.

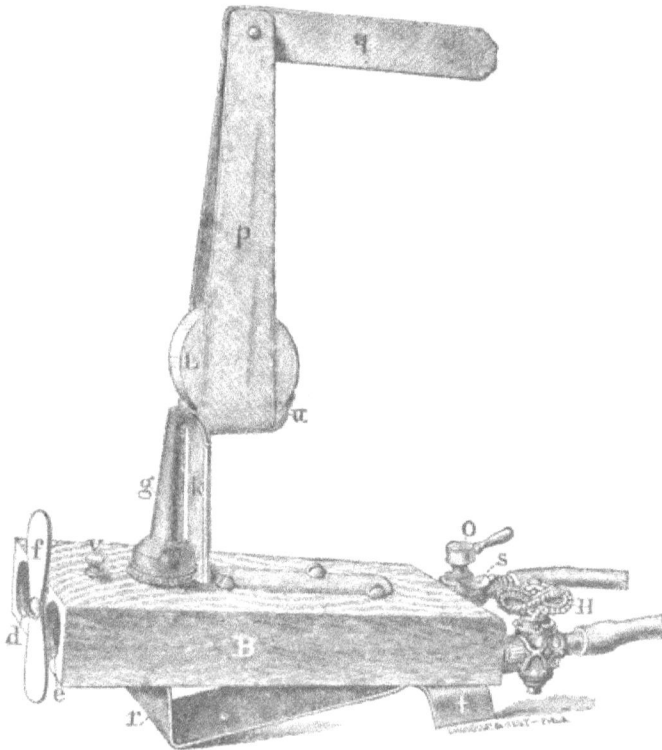

Fig. 40.

Fig. 40 represents the house gas-burner in position,
while Nos. 1 and 3 are in the cells *d* and *e*, to be shut in
by the cover *f*.   The broach in the cell *v* is in proper

shape for entering and clearing the flat mouth of the jet. The wooden base plate, *B*, the bottom of which is shown at Fig. 41, holds the nozzle *n*, which receives into its middle aperture the current of oxygen gas from *O*, and into its side aperture either alcohol or hydrogen gas from *H*. The lime, *L*, in its holder, *P*, is let down the open chimney of the Sciopticon and held in place by the socket *k*, the elongated top of which serves as a guide to the stem, a rest for the lime cradle *u*, and a shield from the alcohol wick. The base-plate of the socket is pushed under the screw-heads, into contact with the nozzle *n*, and its angle and its height are such as to bring the lime disk into proper position. The lime disk is ⅝ of an inch thick, and 2 inches in diameter. It lies

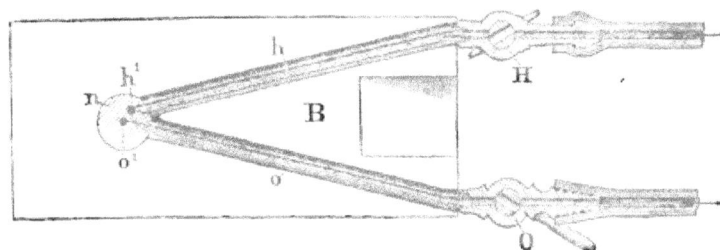

Fig. 41.

loosely in its cradle, *u*, so that lifting the back with the thumb and finger half an inch or so, brings down to the jet a new but hot place in the lime. This operation is not hard on the fingers, because the move is so easily and quickly made, and at a place in the non-conducting lime most distant from the jet.

THE SCIOPTICON WITH THE TRIPLE JET.

If the Sciopticon is required only for the lime light it is furnished without the oil lamp and flame-chamber.

With the Sciopticon complete, as illustrated and described at page 30, we remove the back flame-chamber
glass, G', the narrow strip F, the chimney cap J, and the
oil lamp S. With the lime light the front glass G is
only useful to protect the condenser; if retained, it should
be glass without the cut as commonly used, or else clear
mica. With a concentrated light every blemish shows
upon the screen.

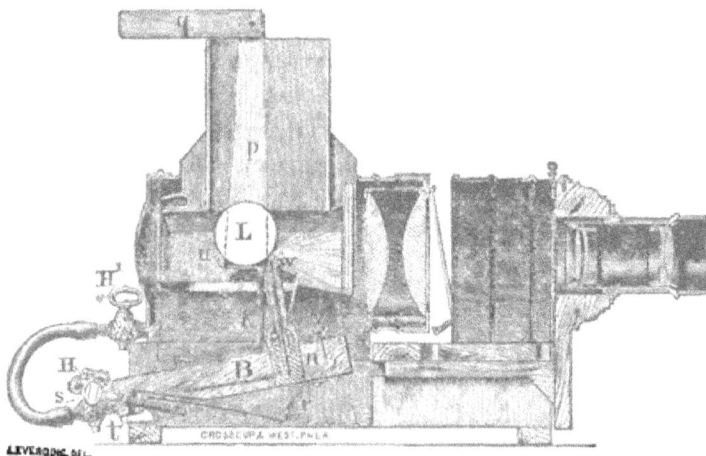

Fig. 42.

Removing the lime-holder from its socket we insert
the blow-pipe by dipping the jet under and into the
flame-chamber, and letting the lime with its holder down
the open chimney into its place again, as shown in section (Fig. 42).

Oxygen gas in a bag between pressure boards weighted
by from fifty to one hundred and fifty pounds, or else
compressed in a condensing cylinder, is put in connection
with the oxygen jet at the stop-cock O, known by its
lever-shaped key. The oxygen should be habitually the
last to be let on, and the first to be shut off. The mouth

of the jet should be kept clear by the broach *v,* and the
tubing should be kept free from kinks.

For the alcohol-burner the wick tube, loosely filled
with wicking, is put in connection with the alcohol
fountain held to the ledge outside by a spring.

For the gas-burner the conical tube is substituted for
the wick tube and connection is made with the house
gas fixtures by sufficient length of flexible tubing.

For the mixed jet connection is made with hydrogen,
in bag or cylinder, subject to the same amount of pressure as the oxygen. House gas collected from the burner
is mostly used, when obtainable, to save the trouble and
expense of generating pure hydrogen.

After lighting the hydrogen (vapor or gas), and allowing it time to heat and dry the lime, the oxygen is
turned on gradually in such force as to produce the best
effect.

The beginner may remove the condenser to see the
light in front, but it is more convenient and easier for
the eyes to observe it from behind by its reflection in
the front glass. The light is shown in the diagram in
about its right place; moving it in and out, and varying
its height by the thumb screw *s,* will readily settle its
right position.

The ordinary objective twenty feet away, throws a
disk of about ten feet in diameter; or in general, its
disk is about half the measure of its distance. To project a picture to a great distance without too much
enlargement, the objective must be of low power and
the extension front must be considerably drawn forward.
The ordinary objective with the back lens removed will
answer, but there is a noticeable want of sharpness
about the margin of the projection. A combination of
low power, made expressly for the purpose, is better.

It is a mistake to suppose that pictures intended for the Sciopticon should be selected with special reference to the light used. The proper density for the Sciopticon flame illumination differs little if any from what is best for the lime light proportionally enlarged. The class of customers most anxious to secure clear, bright pictures are professional exhibitors with a lime light. Dense pictures fogged in the high lights cannot be shown to advantage by any means of illumination.

PECULIAR ADVANTAGES OF THE TRIPLE BLOW-PIPE.

It is mounted in wood, which holds the parts firmly, and in convenient position, and is a non-conductor of heat.

The jet strikes the periphery of the lime disk, which presents larger surface than its side, with less obstruction to escaping heat; strikes it above, so as not to shade the condenser; strikes the lower-half, so the upper part shades the light, instead of a chimney cap; strikes it obliquely, so as not to be reflected back upon the apparatus.

The flat mouth of the jet secures to a fuller flow of gas the proportional efficiency of a smaller opening, and moreover favors the mingling of the gases outside as effectually as it is done in the mixed jet, or nearly so.

The elongated and curved top of the lime-holder socket serves as a rest for the lime cradle, as a guide to its stem, and as a shield from the alcohol wick.

The lime lies loosely in its cradle, easy to turn, resting securely even if broken, and is held to the jet without variableness.

The height of the alcohol in the wick-tube is little affected by tilting the lantern, because the fountain rests abreast of it. The fountain-cup may rest at either side of the lantern.

The three burners are interchangeable with each other, and the triple blow-pipe itself is interchangeable with the oil lamp; so the Sciopticon is furnished for all places and occasions.

THE DISSOLVING COCK.

The oil lamps in a pair of dissolving lanterns are kept steadily burning, while the light of each is cut off from the screen alternately by the crescent-shaped dissolver, as shown Fig. 16, p. 40.

With the blow-pipe the lights themselves are made bright alternately. The expense of keeping two under full head, when only one shines upon the screen at a time becomes worth considering; besides the external cut-off does not produce so soft and pleasing an effect with a concentrated light.

With alcohol burners the oxygen is switched off, so to speak, from one to the other alternately; the deserted lime becoming dim at the same rate its alternate becomes incandescent. The dissolving cock, held to the stand by screw-heads, has a nipple to connect with the oxygen reservoir, and one for each blow-pipe.

With the mixed jet, the flame, when deserted by oxygen, spurts far out, making it necessary to cut off a portion of the hydrogen also. The mixed jet stop-cock therefore has six nipples, three for oxygen and three for hydrogen.

Dissolving views with gas burners also need the double dissolving cock, but one that gives as much more freedom to the hydrogen, as its force is weaker.

A dissolving cock cannot be used in connection with condensing gas cylinders, because the tubing will not stand the pressure. The gas from them must be controlled by stop-cocks at their head.

PREPARATION OF OXYGEN GAS.

MATERIALS.—Theoretically, one pound of chlorate of potash should yield 37 gallons, or 5 cubic feet of oxygen gas; or enough to fill the ordinary 30 by 40 inch rubber bag. In common practice, however, it takes 20 ounces to get 5 feet, or a quarter of a pound to a cubic foot.

To facilitate the decomposition at a lower temperature, and to moderate the flow of gas, we mix with the 20 ounces of chlorate of potash about 5 ounces of black oxide of manganese.

To be assured that this black powder is no part charcoal, black lead, sulphide of antimony, or any thing else that will make with the chlorate of potash an explosive mixture, we may mix and heat a sample of a new supply on a scrap of sheet-iron, or in an iron spoon, over a lamp. If it simply melts and dries away, leaving a dark gray residuum it is safe; if it flashes up, leaving a whitish residuum it is unsafe.

For habitual use, it is convenient to keep this oxygen mixture in stock. Put into a box, say 20 pounds of pure chlorate of potash, broken, so as to pass readily into the retort. Add to these white, broken crystals, 5 pounds of black oxide of manganese, and stir the two well together into a dark gray mass. A pint cup is convenient as the measure of a "charge," as it holds besides the manganese about a pound of the chlorate.

THE APPARATUS.—A gas stove, a (Fig. 43), where we may have it, is, perhaps, the most convenient heating apparatus. An alcohol lamp, as commonly recommended, is too slow, or else with larger wicks it is in danger of explosion. A kitchen stove is better, either in the kettle's place with a brisk fire, or else upon the live coals.

A conical sheet-iron retort, *b*, about a foot high, with joints "up set" and hammered close, is cheaper than the copper retort in common use; is handier, stands firmer, lasts longer, and can be new-bottomed by any tin-smith when burnt out. When new, the seams should be luted with moistened clay or plaster of Paris, and whenever used the cap *c* must be luted on.

Fig. 43.

The cap has the same bevel as the retort, tapering into a bent tube, the end of which is covered by the flexible tubing *d*. The gas when liberated by heat passes through this tubing, first down the long pipe into the water, near the bottom of the wash bottle *e*, then bubbling up, washed and cooled, it passes over and into the gas bag *O*. It will be noticed that connections are made in all our apparatus, by slipping the flexible tubing over the ends of the brass pipes, which either have tapering nipples, or are cut with a slant on the under side.

THE OPERATION OF COLLECTING OXYGEN GAS.—Pour the charge into the retort, seeing that no chips or other materials enter with it. Let the wash bottle be less than half full of water. Lute on the cap with moistened plaster of Paris, and make the connections as shown in the diagram, except that the outlet pipe, f, of the wash-bottle may be left open for a moment or so, until the flow of gas expels the air; see that the stop-cock is open and that the tubing is unobstructed.

Apply sufficient heat to almost immediately melt that portion of the charge in contact with the bottom of the retort, then as the rest melts in turn the operation will be gradual. A slow fire is to be avoided; for it, after a tedious waiting, raises the whole charge to about the melting point, when the decomposition suddenly proceeds with frightful rapidity, perhaps choking the passages and parting the connections. The connections, however, are so easily parted that there will be at the worst only annoyance and loss of gas, but no danger. It is a common recommendation to abate the heat if the flow is too rapid, but with a good heat from the start, the operation is expeditious and safe.

When the bubbling ceases and we conclude from the quantity of gas that the charge is spent, we disconnect the retort and remove it from the fire, and close the stop-cock at the bag.

It rusts the retort less to break up the residuum with a rod, getting it out dry; but it is easier and perhaps better to pour in water and rinse it out, drying the retort directly afterwards.

PREPARATION OF HYDROGEN GAS.

Hydrogen, one of the constituents of water, is produced by decomposing that fluid with zinc and sulphuric acid. A few hours before generating the gas, a mixture of one part, say a pound, of strong sulphuric acid (oil of vitrol) and seven parts of water is made. Consider-

able heat is produced in making the mixture, and for
this reason it should be made beforehand in an earthen-
ware, not a glass vessel, so as to allow sufficient time for
it to become cool before being added to the granulated
zinc contained in the generator *a* (Fig. 44).

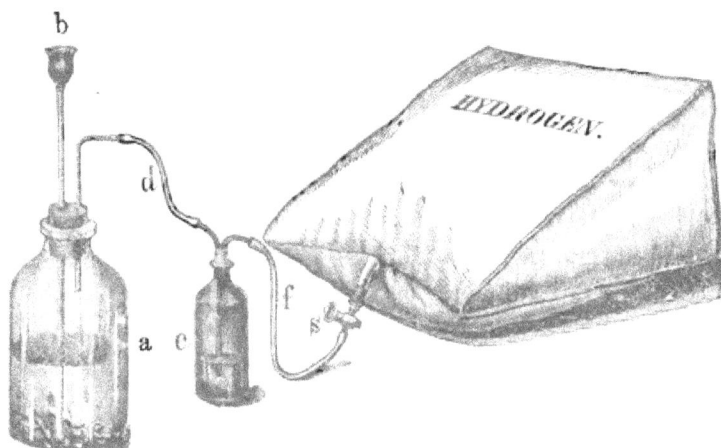

Fig. 44.

About half a pound of zinc is introduced into the
generator (a glass bottle to hold three gallons or more,
or a vessel made of lead), the top of which, bearing the
tube funnel and bent exit tube, is then replaced, and the
joint being made airtight (in the case of a leaden generator by means of a screw, and in the case of a glass
bottle, by a metallic stopper coated with rubber), the
diluted acid is poured down the long tube funnel *b*, the
end of which descends far enough into the liquid to
prevent the return of gas in that direction. A brisk
action ensues, the gas effervescing like so much soda-
water. The first portions should, however, be allowed
to escape for some minutes at the outlet of the wash
bottle *c*, to expel the air. To ascertain when hydrogen
begins to flow, we might apply a light to soap-bubbles

blown from it into a saucer, or to the aperture itself if protected by a fine wire gauze thimble; but such troublesome
precautions rather tend to incur a risk, where there
would be none without them. We can judge near enough
from appearances when to complete the connection.

Where the precaution of diluting the sulphuric acid
and allowing the mixture to cool has been neglected,
and sufficient time cannot be allowed for the purpose,
the zinc and water may be placed in the generator, and
the concentrated acid slowly poured down the tube
funnel as it is required.

It is equally important that, before collecting the gas,
the bag in which it is to be received should be pressed
quite flat, or rolled with the stop-cock open, so as to
exclude all trace of atmospheric air. The time when
pure hydrogen is coming off may be known by the rapid
rise of the bubbles to the top of the water, and by the
accompanying sound, which the ear will recognize, after
a little practice, as being unlike that of other gases.
The purifier e should be about half filled with water;
and connection being made between the exit tube f
and the gas bag by means of india-rubber tubing, as
shown in the cut, be careful to turn on the stop-cock s, in
order that the gas may have free entrance into the bag.

The process here given is the simplest of the several
in common use, and the best for collecting hydrogen
gas in a not very large quantity. The self-condensing
gas cylinder, to be next described, promises to super-
sede the more complicated methods, so that their insertion here would be useless.

It may be proper here to suggest, that after an exhibition the bags, particularly the one marked "II," if
not to be soon used again, should be completely emptied,
not only to preserve them, but to insure having fresh
gas next time.

PRESSURE BOARDS.

Instead of the ordinary iron hinges, which only allow the pressure boards to open from the line of contact, two long leather straps, pierced with holes, may be permanently attached to the lower board and hitched to screw-heads on the upper board, allowing it to be in a plane nearly parallel with the lower board, while the bag of gas is between them and the weight bears on the side opposite. These straps may be hitched up, from time to time, as the gas is expended. By giving sufficient length to these strap-hinges, the two bags for the mixed jet may be placed one upon the other and subjected to the same pressure.

The three boards hinged together in the shape of the letter **Z**, to receive a bag in each angle, as commonly recommended, are not only heavy and expensive, but a measure could hardly be devised more likely to give unequal pressure. A long board extending from one bag to the other, with the weight upon the middle, would be better.

The necessity of exactly equal pressure to be given to the two gases used with the mixed jet, is not so absolute as might be inferred from the way it is usually spoken of. It is surely well to see that the bags are about equally weighted. When two gas cylinders are used, one nearly spent need not be mated with one fully charged. If, however, the pressure in each is in excess of what is needed, the stop-cocks are made to regulate the flow.

As it is inconvenient to transport heavy weights from place to place, traveling exhibitors may fill a box or bag with brick or stone at each place of exhibition. This expedient affords a steadier weight than to seat boys upon the pressure-boards, and more continuous than can be effected by clamping screws.

12

EDGERTON'S SELF-CONDENSING GAS CYLINDERS.

"Special attention is asked to these cylinders, affording as they do a more easy and safe means of producing and condensing the gases for stereopticon purposes and general illumination. To the traveling exhibitor they fur-

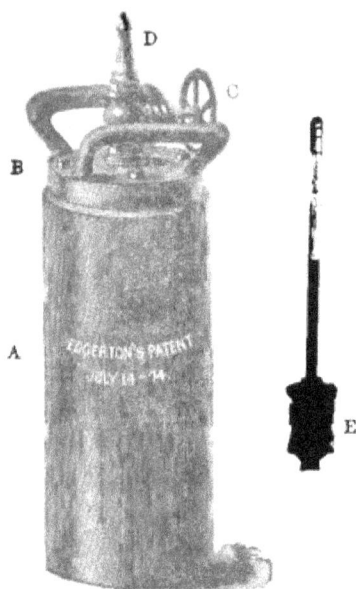

Fig. 45.

nish a compact means of transporting his gases, and save the labor and vexation of carrying weights, pressure-boards, etc.; while to the teacher they are invaluable, placing at his command, at all times, a powerful light as readily started and as easily managed as that of a coal-oil lamp.

"They are made of wrought iron, with a cast-iron cap, and are capable of sustaining a pressure twenty times as great as the strain they are subject to. Referring to the cut, A is the wrought-iron cylinder, B the cast-iron cap, C the valve, D the nipple for hose, and E the pressure-gauge. The hydrogen cylinder is coated with vulcanized rubber and is proof against the action of the sulphuric acid. It is usually a size larger than the oxygen cylinder.

"*To operate the cylinder for oxygen*, unscrew and remove the cap; then set the cylinder over the fire (a range or stove preferred) until quite warm and entirely dry; then pour in the chlorate of potash (one pound) and the black oxide of manganese (four ounces). See that none of the mixture falls upon the cylinder head, so as to prevent the cap fitting closely down. Now rub a little tallow on the cap to make a smooth joint, replace it so that the marks on the cylinder and on the cap will coincide, and screw the nuts down tight; then screw on the gauge and open the valve. Allow the cylinder to remain on the fire until the gas has come off, which will be indicated by the rise of the colored fluid in the gauge. The gas from one pound of chlorate of potash will raise the fluid to within three-fourths of an inch to an inch of the top of the tube. The cylinder ought not to be made red hot in any part. When the gas has come off, set the cylinder away to cool; and after it has become cold, shut the valve, remove the gauge, and screw on the nipple. It is now ready for use at any time, but can remain in the cylinder for months, if not required sooner.

"When the gas is all used up, shut the valve, and let it stay closed until you wish to make a new lot. This will keep the cylinder dry and obviate the necessity of drying over the fire before recharging. Then, when you wish to make fresh gas, unscrew and remove the cap, tapping the end of the wrench with a hammer if the

nuts are hard to start; take a piece of wood, sharpened
at one end, insert it in the cylinder, and break up the
residuum by a few vigorous blows; pour it out and re-
charge without either washing or drying.

" *To operate the cylinder for hydrogen*, unscrew the cap as
before, put in two pounds of scrap zinc, and add a mixture of sulphuric acid and water (four pounds of acid
and four quarts of water). This mixture should be cold
when poured into the cylinder. As soon as the liquid is
poured in, screw down the cap as before, slip a gum tube
on the nipple, and begin to use as soon as there is sufficient pressure, if it is desirable. There is no practical
use for the meter in this case; if used, the red liquid
will stand within about a quarter of an inch of the top
of the tube when the operation is completed.

"As there is an excess of zinc introduced all the acid
will be neutralized. There will be no deterioration of
the gas or injury to the cylinder from long standing.
When the hydrogen is burned up, pour in water to dissolve the sulphate of zinc. This is easily done, and the
cylinder is then rinsed out, and is ready for another
charge."

We believe with the inventor, as above expressed,
that the self-condensing gas cylinders will prove a very
great convenience in the production of the lime light.

It is obvious that the directions as to fitting the cap
upon the cylinder-head must be strictly observed. Any
particles between the meeting surfaces prevents perfect
contact and so will allow the gas to escape.

The oxygen cylinder, owing to its thickness, requires
a longer heating to disengage the gas than the retort
before described. There is, in this case, no outward
current of gas, dust, or foam, to make former directions
applicable only so far as repeated above.

MULTUM IN PARVO.

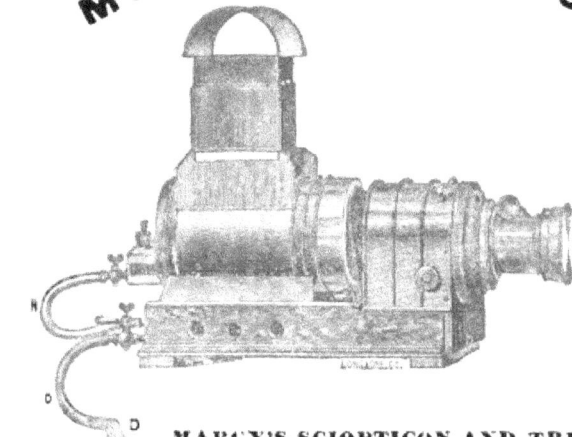

MARCY'S SCIOPTICON AND TRIPLE JET.

Fig. 46.

The condensing gas cylinder occupies but a small portion of the space required by a gas bag with its pressure boards and weights. Considering, moreover, that the apparatus here illustrated gives the best results with comparatively little trouble, the significant heading of "Much in Little," is well-deserved.

With oxygen in the cylinder, or in a bag, we may have the lime light either with alcohol or with gas from house gas fixtures. For the mixed jet, the hydrogen must be forced from a second cylinder or bag.

ATTACHMENTS FOR USE WITH THE LIME LIGHT.

THE LANTERN MICROSCOPE.—This instrument is intended to show natural objects, suitably prepared and
mounted with Canada balsam, between two discs of glass.
They consist of details in the anatomy of a bee, wasp,
flea, spider, larvæ of insects found in stagnant water, as

Fig. 47.

gnats, dragon-flies, parasitic and other insects; parts
of insects, sections of woods, teeth, bones, fossil bones,
shells, lace, silk, muslin, etc.; and as such objects are
smaller than paintings for the lantern, and contain more
delicate details, a proportionately higher magnifying
power is required, which may be adapted to the front
of the Sciopticon.

The ordinary lantern microscope objective, sold at
about $10.00, has a high and low-power combination.
There is, however, all the need of achromatic objectives
for projections that there is for the common microscope;
in which case the cost cannot be less. The apparatus
shown at Fig. 47, with an inch objective, costs about
$60.00. The objectives of a table microscope might
doubtless be adapted to lantern use. Of course there can
be no satisfactory results without proper adaptations,
and perfect alignment and adjustment of distances.

EXPERIMENTS WITH THE LANTERN MICROSCOPE.—By filling a glass trough with diluted sulphuric acid, and dropping into it a few pieces of granulated zinc, the decomposition of water may be shown to an entire audience. Aided by a six-cell Smee's, or Grove's, battery, and a small thin tank, the power which palladium possesses of absorbing nine hundred times its volume of hydrogen may also be shown; the snake-like contortions of the strip of metal, and the bubbles of gas escaping on the reversing of the current, proving very interesting.

The crystallization of salts may also be shown by placing a drop of a strong solution of Epsom salts, or sulphate of copper (blue vitriol), on a piece of glass of suitable size.

Another effective result is obtained by placing in the glass tank a small horse-shoe magnet, and dropping around it some iron filings, which will be found to arrange themselves, or rather be attracted by the magnet, in a most extraordinary manner.

Exhibitions of microscopic objects by the aid of the magic lantern in the drawing-room sometimes fail to give that complete satisfaction which is desirable, owing to attempts being made to show them on too large a scale in proportion to the light employed. We have given very satisfactory exhibitions on a sheet of Imperial (22 x 30) white card-board, fastened by drawing-pins to a board, and fixed against some books or on a chair. In this way the proboscis of a blow-fly may be enlarged to two feet in length, and this is found to be quite large enough for most private assemblies.

It is desirable to have two or three sets of lenses, of different powers, with the microscope, which are varied to suit the object to be exhibited; and it is important to observe that when minute objects are being exhibited,

and a high power consequently in use, the source of
light should be drawn farther from the condensing lenses.
A very interesting addition to the microscope consists
of a diagonal mirror, whereby the image of the objects,
instead of being projected directly on an opaque screen,
may be thrown down at right angles on a sheet of paper
placed on a table, and a drawing very conveniently
made.

HOLMAN'S SIPHON SLIDE (Fig. 48) allows the passage
of a continuous current of water for the purpose of
keeping it cool in the focus of light. It is designed

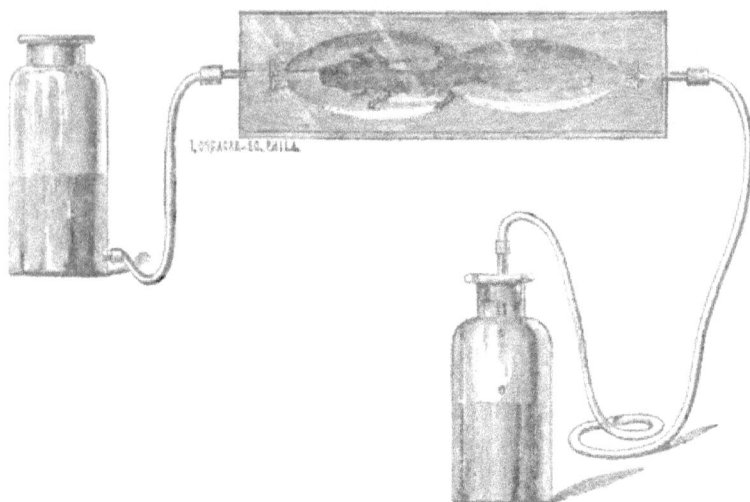

Fig. 48.

for showing the circulation of the blood in a tadpole's
tail, of the sap in plants, &c. Its price, without the
bottles, is $5.00.

A tank filled with a solution of alum is sometimes
used to absorb much of the heat of a beam of light
before it falls upon a delicate microscopic object.

THE MAGIC LANTERN KALEIDOSCOPE.

The KALEIDOSCOPE was invented by Sir David Brewster, in 1814, and all who have witnessed the beautiful effects produced by the instrument will welcome its adaptation to the magic lantern, which, notwithstanding the attendant optical difficulties, has at length been accomplished.

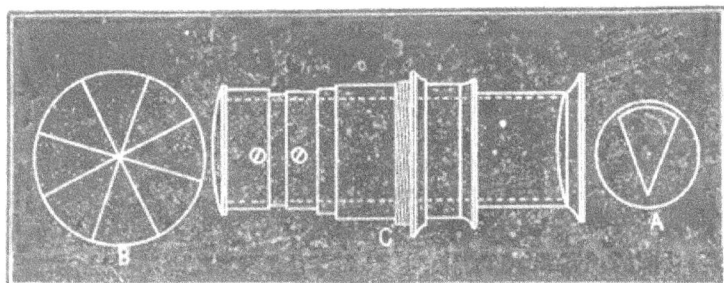

Fig. 49.

The instrument is shown in section at Fig. 49; *A* being a sectional view, showing the disposition of the mirrors; *B*, an outline of the eight-celled image; *C*, a side view of the brass mount, containing the reflectors and lenses, with sliding adjustment for focusing, and projecting the image upon the screen.

It is attached to the lantern by unscrewing the front and screwing the kaleidoscope into its place, turning it round in its sliding tube until the reflectors are upright, like the letter **V**. A rack slide, containing some fragments of colored glass, bugles, beads, and other transparent objects, is also shown; this is introduced into the usual slide-holder of the lantern, and the focus adjusted by sliding the kaleidoscope in or out until its back lens is at a proper distance from the slide.

In a former paragraph, the great importance of having

the various parts of the lantern and the objects to be shown *properly centred* has been dwelt upon at some length. Now, however, the direction is to raise the light about an inch above the centre of the condenser, which can best be done by sliding a narrow board under the blow-pipe. The maximum of illuminating power is obtained in the usual way, by pushing the light backwards and forwards, and the correct focus is obtained by means of the *front* sliding tube. Any dark portions of the image may be removed by turning the kaleidoscope round a very little to the right or left.

The instrument, before using, should be warmed, to prevent what is popularly known as the "steaming of the glass."

Rackwork frames, containing pieces of colored glass, are supplied by the opticians; but exceedingly beautiful effects are obtainable with the chromatrope, a piece of perforated zinc, the bow and the wards of a key, grasses, feathers, a bunch of oats, etc., etc.

THE OXYHYDROGEN POLARISCOPE.

Fig. 50 shows the Oxyhydrogen Polariscope, which consists of two tubes inclined to each other at an angle of 56° 45′, and truncated at their points of junction; the oval space thus formed being closed by some ten or twelve pieces of thin crown glass, the lowest of which is blackened to absorb the polarized ray. This apparatus replaces the object-glass of the lantern, which should have condensers not less than 3½ inches diameter. When attached, it will be seen that the light emanating from the point L, after passing through the condensers C', becomes incident on the crown glass G, inclined at the polarizing angle (56° 45′); the reflected, and in this case polarized, light then passes through the selenite, or other object, in the aperture at O; after which it is brought to a focus by the object-glasses at F, and finally again polarized, or *analyzed*, by the Nicol's prism P, and thence thrown on the screen, the disc on which should not exceed three feet in diameter.

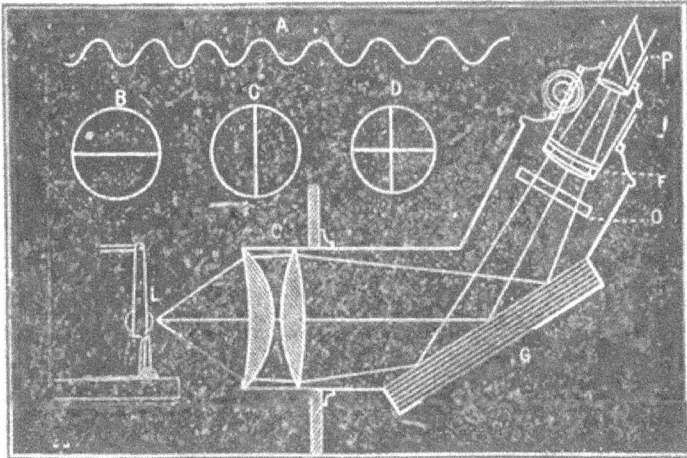

Fig. **50.**

The phenomena connected with the polarization of light are attended by a most gorgeous display of colors, and are, in consequence, among the most attractive in the whole range of physical optics; an apparatus, therefore, which facilitates their exhibition to an audience becomes a most valuable adjunct to the magic lantern.

The subject itself is, however, of too recondite a nature to admit of adequate treatment in the present manual; the reader is therefore referred to Pereira's lectures on "Polarized Light," "Ganot's Physics," and other works on Physical Optics.

The objects best suited for the polariscope are designed with films of selenite of various thicknesses and forms;

sections of quartz, cut in different relation to the axis
of the crystal, producing most splendid tints; unannealed glass, quill, Iceland spar, and, indeed, almost
any matter the particles of which are in a state of
tension. Specimens may be seen, and lists of the various
designs are obtainable, from opticians supplying the
apparatus.

In Fig. 50, the polarizer consists of a bundle of glass
plates, *G*, with the Nicol's prism, *P*, to analyze the
polarized, reflected rays.

Fig. 51.

The polariscope here represented (Fig. 51) consists of
a Foucault prism, of 36 millimetres in diameter, as
polarizer, and a Nicol's prism, of 20 millimetres in
diameter, as analyzer.

THE SCIOPTICON

AND

DISSOLVING VIEW APPARATUS,

WITH A

Priced Catalogue

OF

MAGIC LANTERN SLIDES,

Illustrated and Classified.

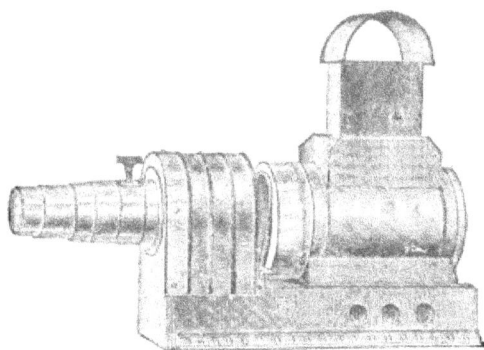

REVISED EDITION.

PHILADELPHIA:

SOLD BY L. J. MARCY, OPTICIAN,

No. 1340 Chestnut Street.

NOTICE.

Slides may be ordered by class and number from the catalogue of any Optician, by giving the name and the edition used.

The receipt of money will be acknowledged by return mail.

When bills are ordered by express, C.O.D., a remittance of ten dollars must accompany the order. The express charge for collection will be added to the amount of the bill.

It will save express charge for collection to send the amount of the bill at once, with the order.

Bills amounting to $100 are subject to a reduction of 5 per cent.

The best mode of remitting money is by a bank draft made payable to my order, or by a post-office money order, or by express.

Making the Sciopticon, and lantern slides, and all appliances in this line a specialty, and having the best of facilities for conducting the business, and of bringing every improvement to bear, I feel assured of being able to fill all orders in a satisfactory manner.

All goods are packed with care, without charge, and are warranted to be in good condition when they leave the store.

Any further particulars that may be desired will be cheerfully given by letter.

Correspondents will oblige by giving in a plain hand the post-office address to which their answers are to be directed, and also the express station to which the goods are to be forwarded.

L. J. MARCY,

1340 Chestnut Street, opposite the U.S. Mint,

PHILADELPHIA.

THE SCIOPTICON

WITH ATTACHMENTS AND ACCOMPANYING APPARATUS.

THE Sciopticon with its oil lamp rather than with its lime light, continues to stand at the head. It is the choice of the many for its being always ready, easy to manage, and inexpensive to use; showing with great brilliancy and steadiness for hours without readjustments or annoyances, and without heating the oil or crack-

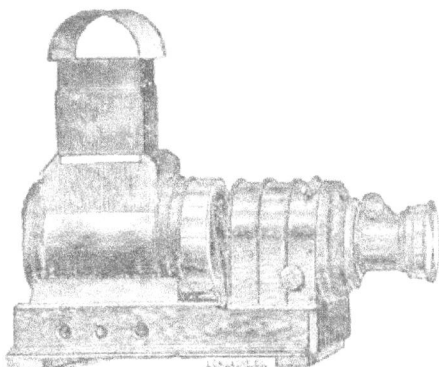

ing the glass; while to close an exhibition we have no further care than to turn down the wicks.

Figures in parenthesis in the following enumeration refer to descriptions and illustrations in the Sciopticon Manual.

1—**The Sciopticon,** with Sciopticon lamp, as illustrated above. With extension chimney; larger opening for escape of heat and for closing the flame-chamber from

(3)

outside; condensing lenses of finest glass, four inches
in the clear, to cover the new large slides (p. 140);
achromatic objective made with special care for the
Sciopticon; opaque curtain operated by milled heads
at the sides; tinters operated by knobs above; stage
with the new stop. All of the latest and most careful
and finished construction, ... $45 00

☞ The new additions and improvements do not add to the
price of the Sciopticon, but they put all further discount out of
the question.

2—**Sciopticon Case,** for carrying the instrument and for
standing it upon when in use, ... 3 00

3—**Sciopticon Case** (same as No. 2), with the addition of
adjustable legs, ... 5 00

4—**Box,** for 100 wooden-mounted slides, ... 2 50

5—**Box,** for 60 wooden-mounted slides, ... 1 50

6—**Double Case,** for a pair of dissolving Sciopticons, which
with its adjustable legs becomes the exhibiting stand;
black walnut, finely finished and polished, ... 10 00

7—**Dissolver,** for a pair of Sciopticons with oil lamps, . 2 00

8—**Pair of** Sciopticons, like No. 1, with case and dissolver, Nos. 6 & 7 (p. 40, Fig. 16), ... 100 00

9—**Sciopticons of Earlier Date,** and lanterns of other
forms, are priced on a sliding scale from $40 to . . 10 00

10—**Marcy's Triple Jet,** for each of the three forms of lime
light (p. 158, Fig. 40), ... 14 00

11—**India-Rubber Bag,** plain, best quality, 30 inches long
by 24 inches wide, with large stop-cock, ... 13 00

12—**India-Rubber Bag,** plain, best quality, 40 inches long
by 30 inches wide, with large stop-cock, ... 16 00

13—**India-Rubber Bag,** cloth-lined, best quality, 40 inches
long by 30 inches wide, with large stop cock, . . . 22 00

14—**Pressure Boards** (p. 169), ... $4 00

15—**Retort Wash-Bottle,** Bag No. 11, connections (p. 165,

    Fig. 43, b, c, d, e, f, s, o), ... 20 00

16—**Gas Stove** (p. 165, Fig. 43, a), ... 2 00

17—**Hydrogen Generator,** copper, ... 15 00

18—**Lime Disks,** two inches in diameter and five-eighths of
    an inch thick, in a sealed can, per dozen, ... 2 00

19—**Oxygen Materials,** in packages having a pound of
    chlorate of potash in each, per dozen, ... 2 00

20—**Zinc,** granulated, per pound, ... 20

21—**Rubber Tubing,** per foot, ... 30

22—**Dissolving Stop-Cock,** for a pair of alcohol lime lights, 6 00

23—**Dissolving Stop-Cock,** for a pair of mixed jet lime
    lights, ... 12 00

24—**Oxygen Self-Condensing Gas Cylinder,** 6 inches in
    diameter and 15 inches high, with gauge and wrench
    (p. 170, Fig. 45), ... 45 00

25—**Hydrogen Self-Condensing Gas Cylinder,** 6 inches in
    diameter and 24 inches high, with gauge and wrench, 55 00

26—**Sciopticon** No. 1, in Case No. 2, ... 48 00

27—**Sciopticon** No. 1, Triple Jet No. 10, in Case No. 2, . 60 00

28—**Gas Bag,** etc., No. 15, with No. 27, ... 80•00

29—**Oxygen Cylinder** No. 24, Sciopticon No. 1, Triple Jet
    No. 10 (p. 173, Fig. 46), net, ... 100 00

30—**Sciopticon with Triple Jet,** but without oil lamp, . 45 00

31—**Kaleidoscope,** 20 00

32—**Microscope** (p. 174, Fig. 46), ... 65 00

33—**Polariscope** (p. 80, Fig. 51), ... 60 00

Chemicals for the tank experiments, chemicals for photographing slides, colors for coloring slides, screens, and all lantern apparatus and appliances not herein enumerated will be furnished at the lowest market prices.

A very convenient Plain Tank prepared specially for the Sciopticon, $1 00
Plain Tank, as in the annexed figure, ... 3 00

Half dozen fine glass plates, 4 inches square, to show crystalliza-
tion, ... 1 50
Pipette, with rubber bulb, to use with tanks, . . . 1 25
Rood's apparatus to show progressive motion of a wave, . . 3 00
Bisulphide of Carbon Prism, to hold in the hand, . . . 5 00
Glass Goblet, to be filled with water in front of condensers, show-
ing at the same time refraction and inversion by lenses, . . 50
Grooved Frame for glass slides, ... 25
Lamp-wicks for the Sciopticon light, per dozen, . . . 25
Flame chamber glasses, per dozen, ... 25

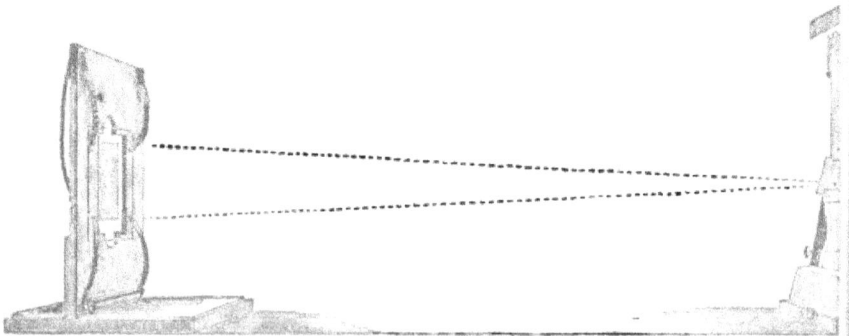

Marcy's Photographic Printing Apparatus, . . 7 00

MAGIC LANTERN SLIDES.

The readiest way of setting forth lantern slides would be to catalogue everything procurable, whether good, bad, or indifferent; so "*you pays your money, and you takes your choice.*" But you find in attractive titles no infallible index to desirable pictures; what you may judge to be a lucky number is not sure to draw a valuable prize.

It is my great desire *not* to sell poor pictures, not only because they fail of giving satisfaction, but because they fail of showing to advantage the merits of the Sciopticon.

The following lists, therefore, are very carefully sifted and arranged, with a view of assisting purchasers in making satisfactory selections. This arrangement applies more particularly to standard colored pictures copied from the great masters, and from scientific illustrations. Plain lantern slides are produced by a rapidly increasing number of photographers, which renders everything like a complete or permanent classification impracticable.

For an exhibition of two hours' duration, forty pictures at least are needed. With a greater number it will require less effort in speaking to entertain the audience.

In cases where it is admissible, two or three verses, every now and then, of some familiar hymn thrown upon the screen relieves the lecturer, and never fails of bringing out the musical talent of the company, and of producing a high degree of satisfaction. When photographed hymns are not obtainable, they may be written on gelatine-covered glass.

Slides can be conveniently ordered by only specifying their class and number.

Under a general direction, a specified list will be suggested by letter to purchasers who request it.

Picture slides now in market are very numerous, after all the great masters, early and late, as found in all the principal picture galleries, any of which can be supplied.

It must be confessed, however, that it is not easy to get pictures of all subjects as good as one could wish. They have all been very carefully examined with a view of selecting the best, and of arranging them so as not to perplex purchasers.

The pictures in the first thirty classes, except such as are said to be painted, or are otherwise designated, are photographed on glass, and beautifully and artistically colored.

Class I—Old Testament Illustrations.

PER SLIDE, $2.50.

1 Adam and Eve in Paradise.	7 The Deluge.
2 The Temptation.	8 Noah's Sacrifice.
3 The First Human Family.	9 Tower of Babel.
4 Death of Abel.	10 Abraham and the Three Angels.
5 Cain Builds the First City.	11 Hagar's Departure.
6 Three Tribes descend from Cain.	12 Hagar in the Wilderness.

8

13 Abraham's Sacrifice.
14 Abraham buries Sarah.
15 The Flight of Lot.
16 Rebekah at the Well.
17 Arrival of Rebekah.
18 Isaac blesses Jacob.
19 Jacob's Dream.
20 Jacob in the House of Laban.
21 Joseph thrown into the Well.
22   " sold to the Midianites.
23   " Bloody Coat shown.
24   " interprets Pharaoh's Dream.
25   " makes himself known.
26   " meets his Father in Goshen.
27   " presents his Father to Pha-
     raoh.
28 Jacob blesses the Sons of Joseph.
29 Jacob blesses his Twelve Sons.
30 Moses Exposed.
31 Moses Saved.
32 The Burning Bush.
33 Pharaoh entreats Moses.
34 Pharaoh and his Host drowned.
35 The Song of Miriam.
36 Gathering Manna.
37 Moses smiting the Rock.
38 The Brazen Serpent.
39 Moses receiving the Tablets.
40 Moses descends from Sinai.
41 Falling Walls of Jericho.
42 Jephthah's Daughter meeting her
    Father.
43 Sacrifice of Jephthah's Daughter.
44 Samson and the Foxes.
45 Samson and Delilah.
46 Samson destroying the Temple.
47 Naomi and Ruth.
48 Boaz and Ruth.
49 Samuel and Eli.
50 Saul and the Witch of Endor.
51 David slaying the Lion.
52 David slaying Goliath.

53 David bringing the Ark.
54 Nathan's Parable.
55 Absalom entangled in the Oak.
56 Judgment of Solomon.
57 The Widow's Oil.
58 Ascent of Elijah.
59 Children in the Fiery Furnace.
60 Captives in Babylon.
61 Daniel in the Lions' Den
62 Esther and Ahasuerus.
63 Esther confounds Haman.
64 Jonah exhorts the Ninevites.
65 Jonah cast into the Sea.
66 Jeremiah on the Ruins of Jeru-
    salem.

PORTRAITS.

67 David.
68 Solomon.
69 Isaiah.
70 Ezekiel.

———

71 Hagar.
72 Ruth.
73 Rachel.
74 Rebekah.
75 The Wife of Potiphar.
76 Pharaoh's Daughter.
77 Deborah.
78 Jephthah's Daughter.
79 Delilah.
80 Hannah.
81 Abigail.
82 Jezebel.
83 The Queen of Sheba.
84 Esther.
85 Athalia.
86 Judith.
87 The Mother of the Maccabees.
88 Sarah, Wife of Tobias.

Class II — New Testament Illustrations.
PER SLIDE, $2.50.

1 The Annunciation.
2 Naming of John the Baptist.
3 The Birth of Christ.
4 Birth of Christ announced to the
    Shepherds.
5 The Star of Bethlehem.
6 The Adoration of the Magi.
7 Presentation in the Temple.

8 Flight into Egypt.
9 Slaughter of the Innocents.
10 Christ disputing with the Doctors.
11 John preaching in the Wilder-
    ness.
12 Baptism of Christ.
13 Calling of Matthew.
14 The Wedding at Cana.

15 Christ and the Samaritan Woman.
16 Christ Preaching on the Sea of Galilee.
17 Christ Healing the Sick.
18 The Sermon on the Mount.
19 Christ Stilling the Storm.
20 Resurrection of the Daughter of Jairus.
21 Christ Walking on the Water.
22 The Transfiguration.
23 The Good Samaritan.
24 The Parable of the Lost Sheep.
25 Lilies of the Field.
26 Christ Healeth the Blind.
27 The Ten Virgins.
28 The Door was Shut.
29 The Unmerciful Servant.
30 The Prodigal Son.
31 Laborers in the Vineyard.
32 The Wicked Husbandmen.
33 Lazarus at the Gate.
34 Pharisee and Publican.
35 Christ Blessing Little Children.
36 The Sick of the Palsy Cured.
37 Resurrection of Lazarus.
38 Christ entering Jerusalem.
39 Mary anointing Jesus' Feet.
40 Christ clearing the Temple.
41 The Tribute Money.
42 The Poor Widow's Two Mites.
43 Predicting the Destruction of Jerusalem.
44 The Last Supper.
45 Washing the Disciples' Feet.
46 Judas' Kiss.
47 Jesus in the Garden of Gethsemane.
48 Jesus before Pilate
49 Peter's Denial.

50 The Flagellation.
51 Christ Crowned with Thorns.
52 Christ Insulted.
53 Christ bearing his Cross.
54 The Crucifixion.
55 Descent from the Cross.
56 Burial of Christ.
57 The Three Marys.
58 Mary Magdalen at the Sepulchre.
59 Christ and the Disciples at Emmaus.
60 Doubting Thomas.
61 Ascension.
62 The Pentecost.
63 Conversion of Saul.
64 Paul at Athens.
65 Paul at Ephesus.
66 St. John's Vision of the Celestial Jerusalem.

PORTRAITS.

67 Our Saviour.
68 Ecce Homo.
69 John the Baptist.
70 St. Matthew.
71 St. Mark.
72 St. Luke.
73 St. John.
74 St. Peter.
75 St. Paul.
76 St. Andrew.
77 St. Stephen.
78 St. Thomas.
79 The Child Christ.
80 Madonna in the Chair.
81 Madonna San Sixtus.
82 Mater Dolorosa.

Class III—Holy Land and Egypt.

PER SLIDE, $2.50.

1 Jerusalem from the Mount of Olives.
2 Enclosure of the Temple Area.
3 Mosque of Omar.
4 Mount Zion from Hill of Evil Council.
5 Mount of Olives from the Wall.
6 Tower of Hippicus.

7 Church of the Holy Sepulchre
8 Jews' Place of Wailing.
9 Arch in Via Dolorosa.
10 The Golden Gate.
11 Garden of Gethsemane.
12 Bethlehem.
13 Etham, near Bethlehem.
14 Fields of Bethany.

15 Hebron.
16 Harem at Hebron.
17 Ancient Masonry near Hebron.
18 Pool at Hebron.
19 Pool at Siloam.
20 Pool of Hezekiah.
21 Solomon's Pool, near Bethlehem.
22 Well and Remains of Pool at Bethel.
23 Well near Emmaus.

24 Well of the Virgin.
25 Lake of Tiberias from Castle Saphet.
26 Baths and City of Tiberias.
27 Nazareth towards Esdraelon.
28 Vale of Nazareth.
29 Well of Nazareth.
30 Ramleh, with the Hills of Judea.
31 Sidon and Mount Lebanon.
32 Hills of Samaria.

JERUSALEM FROM THE MOUNT OF OLIVES.

Jerusalem is situated at an elevation of nearly 2600 feet above the level of the Mediterranean, on the crest of a hill, having a gentle slope towards the Mount of Olives, so that although the latter is only some 150 feet higher than the city, almost every building is distinctly visible from this point. The two eminences are separated by the deep valley of Jehoshaphat, in the basin of which may be seen the Garden of Gethsemane, now inclosed with a stone wall. The most conspicuous building in the city is the Mosque of Omar, which (as well as that of Aksa to its left) occupies the site of Solomon's Temple. This area is 510 yards long by 318 yards broad. Much of it is seen to be planted and adorned with fountains, &c., and serves as a promenade. Admission to this area is now rigidly forbidden to Christians. To the extreme left of the city, and without its walls, is Mount Zion, with the tomb of David, now also a Mohammedan mosque.

33 Mount Hermon.
34 Mount Carmel.
35 Mount Tabor.
36 Plain Er Raheh, Mt. Sinai.
37 Mount Hor.
38 Mount Nebo.
39 The Rock of Moses.
40 Mount Ararat.
41 Sarepta.
42 Rachel's Tomb.
43 Tombs in the Valley of Jehosha-
    phat.
44 Tomb of the Virgin.
45 Halt above the North End of the
    Dead Sea.
46 Damascus.
47 Scene near Ramleh.
48 Arab of the Desert.
49 Arab Camp.
50 Interior of a Caravansera.
51 Range of the Tombs, Petra.
52 Cæsarea.
53 Fords of the Jordan.
54 Tarsus.
55 Falls of the Cydnus.
56 Map of Palestine.
57 Theatre at Ephesus.

58 Ruins of Persepolis.
59 Ruins of Babylon.
60 Ruins of Balbec.

EGYPT.

61 Ferry at Old Cairo.
62 Street in Cairo.
63 The Shadoof.
64 Nile Boat.
65 Pyramids and Sphinx.
66 The Simoon.
67 Approach to Karnac.
68 Karnac.
69 Columns of Grand Hall, Karnac.
70 Colossi of the Plains.
71 Obelisk and Propylon Luxor.
72 Colossal Statue of Rameses the
    Great, at the Memnonum.
73 Approach to Philoe.
74 View from Philoe.
75 Sculptured Gateway, Philoe.
76 Pharaoh's Bed, Philoe.
77 Portico of Tem. Kalabshe, Nubia.
78 Tombs of Memlook Kings, Cairo.
79 Pylon of the Temple of Edfou.
80 Monument of Heliopolis.

Class IV—Ancient Greece and Rome.

PER SLIDE, $2.50.

ANCIENT GREECE.

1 Plan of Athens.
2 Ancient Athens restored.
3 Ruins of Athens.
4 The Pyræus.
5 Mars Hill.
6 The Philosopher's Garden.
7 Ruins of the Parthenon.
8 The Parthenon restored.
9 Temple of Jupiter, at Olympia.
10 Oracle at Delphi.
11 Sacrifice to Neptune.
12 Sacrifice to Mars.
13 Statue of Pallas Athenæ.
14 Olympian Games.
15 Grecian Warriors.
16 Grecian Chariot.
17 Grecian Dwelling (interior).
18 Grecian Ceremony before Mar-
    riage.
19 The Areopagus.
20 The Assembly of the Gods.

ANCIENT ROME.

21 Map of Rome.
22 Ruins of Rome.
23 Trajan's Arch.
24 Roman Cavalry.
25 War Elephant.
26 War Engines.
27 Victorious General thanking his
    Army.
28 Prisoners passing under the Yoke.
29 Roman Triumph.
30 Captives in the Forum.
31 Gladiators at the Theatre.
32 Gladiators at Funerals.
33 Sea Fight.
34 Roman Feast.
35 The Coliseum.
36 Section of Coliseum.
37 Wild Beasts and Victims in the
    Coliseum.
38 Sacrifice in Rome.
39 Temple of the Sun in Rome.
40 Funeral of an Emperor.

Class V—Complete Illustrations to the Text of the Holy Bible.

PER SLIDE, $1.50.

BIBLICAL ANTIQUITIES.

EGYPTIAN.

1 Ancient Cymbals, &c.
2 Ancient Egyptian Armlets.
3 Ancient Egyptian Doors.
4 Ancient Egyptian Scales.
5 Ancient Egyptian Seats.
6 Body of Archers.
7 Bowing before a Public Officer.
8 Brickmaking.
9 Carrying Corn.
10 Chairs.
11 Couches.
12 Culinary Vessels.
13 Dandour.
14 Denderah.
15 Drawers and Girdle.
16 Earrings of Men.
17 Edfou.
18 Egyptian Amulets.
19 Egyptian Entertainment.
20 Egyptian Instrument.
21 Egyptian King on his Throne.
22 Egyptian Lady.
23 Egyptian Lamps.
24 Egyptian Vessels of Elegant Form
25 Egyptian with a Tray of Meats.
26 Ephod and Censer.
27 Ephod and Girdle.
28 Ethiopian Car drawn by Oxen.
29 Fauteuils.
30 Harp.
31 Luxor.
32 Man-servant.
33 Metal Door-pins.
34 Mitres.
35 Mummy.
36 Mummy Case and Marble Sarcophagi.
37 Overseer of Cattle.
38 Priestesses.
39 Ring Money.
40 Rock-cut Temple, Ipsambul.
41 Scarabæi—Back and Side Views.
42 Scarabæi—Engraved under-surfaces.

43 Scribe.
44 Ship.
45 Side View of Memnon.
46 Signet Rings of Ancient Egypt.
47 Sistrums.
48 Sphinx and Pyramids.
49 Statue of Egyptian Lady
50 Stewards.
51 Stringed Instruments.
52 Tambourine Players.
53 Theban Statue.
54 Thrones.
55 Water Bearers.
56 Windows.
57 Wine-press.
58 Worker in Iron.

JEWISH.

59 Altar of Burnt Offering.
60 Altar of Incense.
61 Booths.
62 Burnt Offering.
63 Costume of High Priest.
64 Costume of a Priest.
65 Meat Offering.
66 Ox-horn Blower.
67 Peace Offering.
68 Priests Sounding an Alarm.
69 Setting up the Tabernacle.
70 Sin Offering.
71 Solomon's Throne.
72 Supposed Form of the Laver.
73 Table of Shew-Bread.
74 The Golden Candlestick.
75 The Princes' Offering.
76 Trespass Offering of the Poor.
77 Alabaster Boxes.
78 Alexander the Great.
79 Ancient Battering Ram.
80 Ancient Shadoof.
81 A Phœnician Sarcophagus.
82 Ark borne by Priests.
83 Balista prepared for the Discharge of a Stone.
84 Bas-relief from the Arch of Titus.
85 Captive Jews.

86 Catapulta prepared for the Discharge of an Arrow.
87 Censers.
88 Chamber on the Wall.
89 Coin of Agrippa (Copper).
90 Coin of Archelaus (Copper).
91 Coin of Augustus.
92 Coin of Claudius.
93 Coin of Nero.
94 Coin of Tiberius.
95 Coin of Titus.
96 Cuirass.
97 Daggers.
98 Demi-Shekel (Copper).
99 Double Flutes, Greek.
100 Eastern Tables.
101 Eastern Writing Material.
102 Escape from a Window.
103 Garden Bedstead.
104 Garden House.
105 Gods of Wood.
106 Group of Altars.
107 Helmets.
108 Insignia of Office.
109 Interior of the Portico of the Great Temple of Denderah.
110 Judæa Capta.
111 King on Throne, with Attendants.
112 Metal Mirrors.
113 Nimrod.
114 Nisroch.
115 Palm Bedstead.
116 Persian Armlets.
117 Persian Torch and Lantern.
118 Pillows of Stone and Wood.
119 Quarter-Shekel (Copper).
120 Roman Judgment-Seat.
121 Roman Lantern and Flambeaux.
122 Sandals.
123 Shekel of Copper.
124 Shekel of Silver.
125 Ship, from a Painting at Pompeii.
126 Shrine with Idol.
127 Spoons.
128 Statue of Cyrus.
129 Sun Dial.
130 Teraphim.
131 The Ckumarah.
132 The Great King.
133 Tower in the Desert.
134 Winged Human-headed Bull.
135 Writing Materials.

MANNERS AND CUSTOMS.

136 A Musical Procession.
137 An Oriental Migration.
138 Application to a Santon.
139 Arab Encampment.
140 Arab Female.
141 Arab Horde coming to a Halt.
142 Bargaining for a Slave.
143 Caravansera.
144 City Gate.
145 Cup Bearers.
146 Dance with Timbrels.
147 Drawing Water from the Nile.
148 Eastern Forms of Obeisance.
149 Eastern House.
150 Eastern Housetops.
151 Eastern Potter.
152 Eastern Prince.
153 Eastern Princess.
154 Egyptian Foot Soldiers.
155 Egyptian Soldiers.
156 Egyptian War Chariots.
157 Egyptian Worship.
158 Feast of Passover.
159 Feast of Tabernacles.
160 Female Mourners at Tomb.
161 Giving Water from Leathern Bottles.
162 Grecian Warrior in Armor.
163 Greek Worshipping with Head Uncovered.
164 Hand-mill.
165 Interior of the Tomb of the Kings at Jerusalem.
166 Jewish Physician.
167 Lady with Face-veil.
168 Marriage Procession of a Bride.
169 Marriage Procession of a Bridegroom.
170 Market at Gate.
171 Monumental Pillars.
172 Mourner at Tomb.
173 Musical Entertainment.
174 Oriental Cart.
175 Oriental Shepherds.
176 Ornaments of Egyptian Females.
177 Painted Eyes.
178 Potter at Work.
179 Pouring Wine from Leathern Bottle.
180 Praying with the Head Covered.
181 Raising Water.

182 Raising Water by the Ckutweb.
183 Reading the Law.
184 Rock-cut Tomb.
185 Roman Centurion.
186 Roman Consul.
187 Roman Eagle.
188 Roman Lictor.
189 Roman Soldiers.
190 Sackcloth.
191 Saddled Asses.
192 Searching for Leaven.
193 Shaving the Head.
194 Sheepfold.
195 Stones of Memorial.
196 The Taboot.
197 Threshing by Animals.
198 Threshing by the Drag.
199 Threshing by the Sledge.
200 Throwing a Javelin.
201 Walking Wrapper.
202 Warrior and Armor-Bearer.
203 Washing Hands.
204 Water Carriers.
205 Women of Priestly Families.
206 Women on Camels.
207 Women Wearing the Tob.

BIBLICAL SCENERY.

208 Absalom's Tomb.
209 Aceldama.
210 Adjeroud.
211 Alexandria.
212 Amphitheatre near Tiberias.
213 Anathoth.
214 Antioch.
215 Aqueduct of Jericho.
216 Arch of Titus, Rome.
217 Ascalon.
218 Ashdod.
219 Assyrian Grave Tower, Lebanon.
220 A View of Petra in Wady Mousa.
221 Baalbec.
222 Banias.
223 Bazaar in Damascus.
224 Bazaar in Jaffa.
225 Bethany.
226 Bethel.
227 Bethlehem.
228 Beyrout.
229 Birs Nimrod, Babylon.
230 Bringing First-fruits to Jerusalem.

231 Carmel.
232 Castle of Sion.
233 Cave at Benias.
234 Caves in the Cliffs of Wady Mousa, Mount Seir.
235 Cave under the Temple Hill.
236 Cedars of Lebanon.
237 Chapel of the Burning Bush.
238 Church of the Holy Sepulchre, Jerusalem.
239 Church of the Nativity.
240 Church Ruin at El Bire.
241 Citadel on Site of Ft. Antonio.
242 Colosse.
243 Corinth.
244 Damascus.
245 Daniel's Grave at Susa.
246 Egyptian Monuments.
247 Egyptian Temple.
248 Elath.
249 Elias's Grotto on Mt. Carmel.
250 Emmaus.
251 Garden of Gethsemane.
252 Gaza.
253 Gibea.
254 Gibeon.
255 Graveyard in Sidon.
256 Grave of Joseph of Arimathea.
257 Harvest in Palestine.
258 Hebron, with the Grave of Machpelah.
259 Hermon, from Meromsee.
260 Interior of Coliseum, Rome.
261 Interior of Convent, Mar Saba.
262 Interior of the Holy Sepulchre.
263 Isaiah's Grave.
264 Jacob's Bridge.
265 Jaffa.
266 Jaffa Gate, Jerusalem.
267 Jerusalem, from Scopus.
268 Jerusalem from the North.
269 Jews' Place of Wailing, Jerusalem.
270 Jews' Quarter, Jerusalem.
271 Jezreel.
272 Joseph's Grave.
273 Kaipha.
274 Kirjath-jearim.
275 Lake of Gennesaret.
276 Laodicea.
277 Lydda.
278 Magdala.
279 Malta.
280 Mars' Hill, Athens.

281 Mosque of Omar, Jerusalem.
282 Mount Ararat.
283 Mount Hor.
284 Mount of Olives from Jerusalem.
285 Mounts Ebal and Gerizim.
286 Mount Serbal.
287 Mount Tabor.
288 Nablous, the Ancient Shechem.
289 Nain.
290 Nazareth.
291 Nineveh.
292 Noah's Grave in Armenia.
293 Noph.
294 Old Jewish Tower.
295 On, or Heliopolis.
296 Patmos.
297 Pergamos.
298 Pharaoh's Palace.
299 Philadelphia.
300 Philip's Well.
301 Pilgrim's Pool, Succoth.
302 Plain of Jericho.
303 Pool of Bethesda.
304 Pool of Gihon.
305 Pool of Hezekiah.
306 Pool of Siloam.
307 Pools of Solomon.
308 Ptolemais.
309 Rachel's Grave.
310 Rama (Arimathea).
311 Remains of Ancient Temple
      Bridge.
312 Rhodes.
313 River Jobbok.
314 Rock of Moses.
315 Rock Valley in the Vicinity of
      Petra.
316 Roman Bridge, Lysanias.
317 Rome.
318 Ruins of Ammon.
319 Ruins of Cæsarea in Palestine.
320 Ruins of Gadara.
321 Ruins of the Forum at Rome.
322 Ruins of the Palace of Nero,
      Rome.
323 Ruins of Tyre.
324 Safed.
325 Samaria, Sebaste.
326 Samaritan Synagogue.
327 Sardis.
328 Sarepta.
329 Shiloh.
330 Sidon.
331 Smyrna.

332 St. John's Hospital, Acre.
333 St. Peter's, Rome.
334 Suez.
335 Summit of Mount Sinai.
336 Tadmor, Palmyra.
337 Tarsus.
338 Terrace Cultivation.
339 The Coliseum, Rome.
340 The Dead Sea.
341 The Grave of David.
342 The Holy Sepulchre.
343 The Jordan leaving the Lake of
      Tiberias.
344 The Mamertine Prison, Rome.
345 The Mujelibe, Babylon.
346 The River Jordan.
347 The Written Rocks, Wady Mo-
      katteb.
348 Thyatira.
349 Tiberias.
350 Tomb at Petra.
351 Tomb of Ezra.
352 Tomb of Mordecai.
353 Tomb of the Kings.
354 Tower of David, Jerusalem.
355 Tower of St. Paul in Damascus.
356 Tyre.
357 Urfah, supposed Ur of the
      Chaldees.
358 Valley and Convent of Sinai.
359 Valley of Gihon.
360 Valley of Jehoshaphat.
361 Vaults beneath Solomon's Tem-
      ple.
362 Vestibule within the Golden
      Gate.
363 Vivia Dolorosa.
364 View in the Land of Moab.
365 View of a Portion of the Ruins
      of Petra.
366 View on the Euphrates.
367 View on the Nile.
368 Watered Garden.
369 Well at Cana.
370 Well of the Virgin.
371 Wells of Moses

BIBLICAL NATURAL HISTORY.

BEASTS.

372 Asses
373 Bat.
374 Bear.
375 Beaver.

376 Camels.
377 Chameleon.
378 Common Dormouse.
379 Coneys.
380 Dark-banded Jerboa.
381 Dogs.
382 Dromedary.
383 Egyptian Fox.
384 Elephant.
385 Four-horned Ram.
386 Gazelles.
387 Greyhound.
388 Hippopotamus.
389 Jackals.
390 Lion.
391 Lioness and Whelps
392 Onyx.
393 Sheep.
394 Syrian Leopard.
395 Syrian Ox, Camel, and Ass.
396 Wild Ass.
397 Wolf.

BIRDS.

398 Bee-eater.
399 Collared Turtle.
400 Cormorant.
401 Hawk.
402 Heron.
403 Hoopoe.
404 Ibis.
405 Osprey.
406 Owl.
407 Partridge.
408 Pelican.
409 Quail.
410 Sea Swallow.
411 Shoveller.
412 Stork.
413 Swallow of Palestine.
414 Syrian Dove.
415 The Aquiline Vulture.
416 The Crane.
417 The Eagle.
418 The Flamingo.
419 The Katta.
420 The Ostrich.

REPTILES AND INSECTS.

421 Crocodile.
422 Egyptian Frogs.
423 Emperor Boa.

424 Hornet.
425 Lacerta Gecko.
426 Lacerta Sincus.
427 Lacerta Stellio.
428 Locust.
429 Scorpion.

TREES AND PLANTS.

430 Almond Tree.
431 Apples of Sodom.
432 Balm of Gilead.
433 Bitter Cucumber.
434 Black Fig Tree.
435 Box Tree.
436 Cactus.
437 Carob.
438 Cinnamon.
439 Cluster of Dates.
440 Cone of the Pine.
441 Coriander.
442 Cypress.
443 Darnel.
444 Date Palm.
445 Dourra.
446 Ears of Wheat.
447 Figs.
448 Fig Leaves.
449 Fitches.
450 Frankincense.
451 Gopher Tree.
452 Gourd.
453 Grapes.
454 Hennah Plant.
455 Holy Bramble.
456 Husks.
457 Hyssop.
458 Jasmine.
459 Jonah's Gourd.
460 Juniper.
461 Lentiles.
462 Lily.
463 Mandrakes.
464 Mustard.
465 Nuts.
466 Oleander.
467 Olive Branch with Fruit.
468 Olive Tree.
469 Orange Tree.
470 Palm Tree.
471 Plane Tree.
472 Pomegranate.
473 Prickly Oak.
474 Reeds.

475 Rice.
476 Rose of Jericho.
477 Rose of Sharon.
478 Sea-goose Foot.
479 Sesamum Oriental.
480 Stone Pine Tree.
481 Strawberry Tree.
482 Sycamore.
483 Sycamore Figs.
484 Tamarisk Tree.
485 Terebinth.
486 Thorn.
487 White Mulberry Tree.
488 Wormwood.

MAPS, &c.

489 Canaan in possession of the Twelve Tribes.
490 Geography of the Hebrews.
491 Journeys of the Children of Israel.
492 Plan of Jerusalem, Ancient.
493 Plan of Jerusalem, Modern.
494 Palestine, to illustrate the New Testament.
495 Travels of St. Paul in Asia and Europe.

Class VI—Views of Interest in different parts of the World.

PER SLIDE, $2.50.

ENGLAND.

1 The New House of Parliament, London.
2 Windsor Castle.
3 The Horse Guards.
4 Greenwich Hospital.

SCOTLAND.

5 Interior of Holyrood Chapel.
6 Melrose Abbey.
7 Balmoral Castle.
8 Fountain's Abbey.

IRELAND.

9 Askeaton Abbey.
10 Adare   "
11 Furness   "
12 Muckross   "
13 Castle and Town of Glenarm.
14 The Custom House at Limerick.
15 Thodmongate Bridge,   "
16 The Coleraine Salmon Lake.
17 Dunluce Castle, County Antrim.
18 View of Londonderry.
19 Walker's Monument, Londonderry.
20 The Green Linen Market, Belfast.
21 High Street,   "
22 Black Rock Castle, County Cork.
23 Statue of George II,   "
24 Cove Harbor,   "
25 Cork River.

26 Merchant Quay, Cork.
27 Blarney Castle.
28 Trinity College, Dublin.
29 College Street,   "
30 Parliament Square,   "
31 The King's Bridge,   "
32 St. Peter's Chapel,   "
33 The Four Courts,   "
34 The Bank of Ireland.
35 Court-Yard, Dublin Castle.
36 Castle Kilkenny, County Dublin.
37 Belfry and Church of Swords, near Dublin.
38 Dunmore Pier, Waterford.
39 Lismore Castle,   "
40 Inchmore Castle, Co. Kilkenny.
41 Kilkenny Castle.
42 The Old Abbey at Sligo.
43 The Boyne Water.
44 The Upper Lake of Killarney.
45 The Lower Lake of Killarney.
46 Innisfallen.
47 Queenstown Harbor.
48 Carric Fergus Castle.
49 The Seven Churches of Clonmacnoise.
50 Abbey of the Holy Cross, Tipperary.
51 Ballanahinels.
52 Enchanted Isles.
53 Giant's Causeway.
54 Cave at Giant's Causeway.
55 Fingal's Cave.

AMERICA.

209 Cape Horn.
210 The Dome, Yosemite Valley, Cal.
211 The Three Brothers, Yosemite Valley, California.
212 Cathedral Spires, Yosemite Valley, California.
213 The Vernal Falls, Yosemite Valley, California.
214 Bridal Veil Falls, Yosemite Valley, California.
215 Grizzly Giant Tree.
216 Pacific R. R. Track on the Rocky Mountains.
217 The Rio Grande, near Frontera.
218 Monument Mountains, Rocky Mountains.
219 The Organ Mountains, New Mexico.
220 Spanish Peaks, New Mexico.
221 Fort Laramie.
222 The Wind River Mountains.
223 Fort Smith, Arkansas.
224 Pend d'Oreille Mission, Rocky Mountains.
225 The Garden of the Gods, Rocky Mountains.
226 Fight with a Grizzly Bear in the Rocky Mountains.
227 Ball-play Dance of Camanche Indians.
228 Archery of Mandan Indians.
229 A Buffalo Hunt—Surround.
230 Camanche Indians at Ball-playing.
231 Buffalo Hunt—The Chase.

232 Three Camanche Indians—Ball-players.
233 Buffalo Hunt—the Near Chase.
234 Wild Horses at Play.
235 Antelope Shooting—the Ambuscade.
236 Encampment on the Plains during a "Norther."
237 The War Dance of Camanche Indians.
238 Encampment surprised by Indians.
239 White Wolves attacking a Buffalo Bull.
240 Catching the Wild Horse.
241 A Buffalo Chasing Back.
242 The Capitol, Washington, D. C.
243 The White House, " "
244 U. S. Treasury Building, Washington, D. C.
245 U. S. General Post Office.
246 U. S. Patent Office.
247 General View of Washington from the Capitol.
248 Washington's Tomb, Mt. Vernon, Va.
249 Washington's Residence, Mt. Vernon, Va.
250 View up Broadway from Herald Office.
251 West Point from Garrison's.
252 General View of Niagara Falls.
253 William Penn's Cottage, Letitia Court, Philadelphia, from old engravings.
254 Continental Hotel, Phila.
255 Independence Hall, "

Class VII—Illustrations of Important Events in American History.

PER SLIDE, $2.50.

1 Landing of Columbus.
2 Marriage of Pocahontas.
3 Embarkation of the Pilgrim Fathers.
4 English Puritans Escaping to America.
5 Landing of the Pilgrims.
6 Landing of Hendrick Hudson.
7 Landing of Roger Williams.
8 Elliott, the first Indian Missionary

9 William Penn Treating with the Indians.
10 Washington Raising the British Flag at Fort Duquesne.
11 Patrick Henry in the Virginia Assembly.
12 Washington, Henry, and Pendleton going to the first Congress.
13 First Prayer in Congress.
14 Surprise of Ticonderoga.
15 Washington leaving for the Army.

16 Washington taking Command of the Army, 1775.
17 Putnam leaving the Plough.
18 Battle of Bunker Hill, 1775.
19 Drafting of the Declaration of Independence.
20 Declaration of Independence.
21 Surrender of Burgoyne.
22 Putnam's Escape.
23 Washington cross'g the Delaware.
24 Treason of Arnold.
25 Capture of Major Andre.
26 Surrender at Yorktown.
27 Com. Perry at the Battle of Lake Erie.
28 General Taylor at Buena Vista.
29 General Scott at Contreras.
30 Bombardment of Fort Sumter.
31 Soldier's Dream.
32 Monitor driving the Merrimac.
33 Picket duty on the Potomac.
34 The First Reading of the Emancipation Proclamation. By Carpenter.
35 Gen. Sherman entering Savan'h.
36 Union Army enters Petersburg.
37 Assassination of Pres't Lincoln.
38 Apotheosis of Abraham Lincoln.
39 American Eagle.

PORTRAITS OF DISTINGUISHED

AMERICANS.

40 George Washington, by Stuart.
41   "   "   by Peale.
42 Martha Washington.
43 John Adams,   2d Pres. U. S.
44 Thomas Jefferson, 3d   "   "
45 James Madison,   4th "   "
46 James Monroe,   5th "   "

47 John Q. Adams, 6th Pres. U. S.
48 Andrew Jackson, 7th   "
49 Martin Van Buren, 8th   "
50 General Harrison, 9th   "
51 John Tyler, 10th   "
52 James K. Polk, 11th   "
53 Zachary Taylor, 12th   "
54 Millard Fillmore, 13th   "
55 Franklin Pierce, 14th   "
56 James Buchanan, 15th   "
57 Abraham Lincoln, 16th   "
58 Andrew Johnson, 17th   "
59 U. S. Grant, 18th   "
60 Lincoln at Home—a beautiful picture of President Lincoln and his son Thaddeus.
61 Henry Clay.
62 Daniel Webster.
63 Stephen A. Douglass.
64 Edward Everett.
65 Washington Irving.
66 Professor Longfellow.
67 Dr. Kane, the Arctic Explorer.
68 Rev. Henry Ward Beecher, D.D.
69 Rev. H. W. Bellows.
70 Hon. W. L. Dayton.
71 Horace Greeley.
72 Hon. John P. Hale.
73 Hon. Hannibal Hamlin.
74 Geo. D. Prentiss, of Kentucky
75 Stephen Girard.
76 Hon. Charles Sumner.
77 Hon. Simon Cameron.
78 Hon. S. P. Chase, Chief Justice U. S.
79 Gov. Andrew, of Mass.
80 Gov. Brownlow, of Tenn.
81 William Penn.
82 Benjamin Franklin.
83 Lewis Cass.
84 Thomas H. Benton.

Class VIII—The Principal Battles of the Franco-German War of 1870.

Drawn on the Spot by S. Kaim, and also the Portraits from Life of the principal actors during the War.

PER SLIDE, $2.50.

1 The Battle of Weissenburg, Aug. 4, 1870.
2 The Battle of Wœrth, and flight of McMahon, Aug. 6, 1870.
3 Storming of Weissenburg, Aug. 4, 1870.
4 General view of the Battle of Sedan, Sept. 1, 1870.

5 On the evening of the Battle of
Sedan, General Reille bearing
surrender.
6 Interview between Napoleon and
Bismarck at Doucheri, Sept. 1,
1870.
7 Napoleon surrendering his sword
to King William, Sept. 2, 1870.

PORTRAITS.

8 King William of Prussia.
9 Queen of Prussia.
10 Crown Prince of Prussia.
11 Count Bismarck.
12 Leopold of Hohenzollern.

13 Von Moltke.
14 Minister of War, Von Roon.
15 General Blumenthal.
16 General Falkenstein.
17 General Von Göben.
18 Napoleon III.
19 Empress Eugenie.
20 Prince Imperial.
21 General MacMahon.
22 General Canrobert.
23 General Frossard.
24 General Bazaine.
25 Garibaldi.
26 Prince Napoleon.
27 General Uhlrick.
28 General Steinmetz.

Class IX—Views in Sets, conveying Moral Lessons.

PER SLIDE, $2.50.

1. THE PILGRIM'S PROGRESS.

A Sunday-school Concert Exercise, called
"The Song of the Pilgrimage," is much used
with this set. Price, 60 cents per dozen.

1 The Pilgrim and his Burden.
2 The Pilgrim at the Gate.
3 The Slough of Despond.
4 Christian and the 3 Shining ones.
5 The Shining Light.
6 The Pilgrim and the Lions.
7 Christian and the Shepherds.
8 The Pilgrims found Sleeping.
9 Vanity Fair.
10 Giant Pope.
11 Christian Arming.
12 Passing through the Waters.

☞ This set can be extended to twenty
pictures, if preferred.

2. CHRISTIANA AND HER CHILDREN.

A Sunday-school Concert Exercise, called
"Christiana and her Children," is much used
with this set. Price, 96 cents per dozen.

1 Christiana and her Children.
2 The Letter.
3 The Man with the Muck Rake.
4 The Bath of Sanctification.
5 Great Heart and the Pilgrims.
6 Fight between Great Heart and
Grim.

7 The Young Pilgrims catechized
by Prudence.
8 The Shepherd Boy in the Valley
of Humiliation.
9 The Pilgrims in the Valley of
the Shadow of Death.
10 The Pilgrims at the House of
Gaius.
11 Death of Giant Despair.
12 Christiana preparing to Cross the
River.

☞ Instead of this, a new Series of twenty
slides can be furnished, if desired.

3. DRUNKARD'S PROGRESS AND END.

1 Domestic Happiness.
2 The Temptation.
3 A Loving Heart made Sad.
4 The Rum-hole—a Substitute for
Home.
5 Rum instead of Reason.
6 Degraded Humanity.
7 The Cold Shoulder by old friends
8 Rumseller's Gratitude.
9 Poverty and Want.
10 Robbery and Murder.
11 Mania-a-potu—the Horror of
Horrors.
12 The Death that precedes Eternal
Death.

In Series 3 we just get a glimpse at the
comic side of the melancholy career more
than in Series 4.

4. THE BOTTLE.

From the originals by G. Cruikshank.

1 The bottle is brought out for the first time. The husband induces his wife "just to take a drop."
2 He is discharged from his employment for drunkenness. "They pawn their clothes to supply the bottle."
3 An execution sweeps off the greater part of their furniture. "They comfort themselves with the bottle."
4 Unable to obtain employment, they are driven by poverty into the streets to beg, and by this means still supply the bottle.
5 Cold, misery, and want destroy their youngest child. "They console themselves with the bottle."
6 Fearful quarrels and brutal violence are the natural consequences of the frequent use of the bottle.
7 The husband, in a furious state of drunkenness, kills his wife with the instrument of all their misery.
8 The bottle has done its work—it has destroyed the infant and the mother; it has brought the son and daughter to vice and to the streets, and has left the father a hopeless maniac.

5. THE BOTTLE—(COMIC.)

EIGHT SLIDES, PER SET, $12.00.

1 The Toast.
2 Various Brands.
3 Irish Whisky, Scotch Gin, and Five Points Rum.
4 Stabbing.
5 Shooting.
6 Robbing.
7 Hanging.
8 The End.

6. STOMACH OF THE DRUNKARD IN ITS DIFFERENT STAGES OF DISEASE.

APPEARANCE OF THE STOMACH.

1 Of a Temperance Man.
2 Of the Moderate Drinker.
3 Of the Drunkard.
4 After a Debauch.
5 Of a Hard Drinker.
6 Of a Habitual Drunkard.
7 Of a Drunkard on the verge of the grave.
8 During Delirium Tremens.

7. THE GAMBLER'S CAREER.

1 The first seed of the passion planted in the young mind.
2 The development of the passion with higher stakes.
3 Finding himself always the loser, he resorts to false play.
4 He is detected and roughly handled by his friends.
5 Having finally lost his all, he leaves the gambling-house in despair and madness.
6 He ends his life in a mad-house, still occupied with his ruling passion.

8. THE TEN COMMANDMENTS.

ILLUSTRATED IN 12 PICTURES.

1 Thou shalt have no other gods before me.
2 Thou shalt not make unto thee any graven image.
3 Thou shalt not take the name of the Lord thy God in vain.
4 Remember the Sabbath day to keep it holy.
5 Honor thy father and thy mother.
6 Thou shalt not kill.
7 Thou shalt not commit adultery.
8 Thou shalt not steal.
9 Thou shalt not bear false witness against thy neighbor.
10 Thou shalt not covet.
11 Moses receiving the Tables of the law.
12 Moses delivering the Tables of the law to the people.

9. THE LORD'S PRAYER.

ILLUSTRATED.

Original designs by Nisle.

1 "Our Father which art in heaven."
2 "Thy will be done on earth as it is in heaven."
3 "Give us this day our daily bread."
4 "Forgive us our debts as we forgive our debtors."
5 "Lead us not into temptation."
6 "Deliver us from evil."
7 "Thine is the kingdom, and the power, and the glory, forever. Amen."

10. FOUR SCENES FROM THE LIFE OF A COUNTRY BOY.

1 Leaving Home.
2 Temptation and Fall.
3 Farther on—Gambling.
4 At Last—the Forged Check.

11. SHAKSPEARE'S SEVEN AGES OF MAN.

1 The Infant.
2 The School-boy.
3 The Lover.
4 The Soldier.
5 The Justice.
6 The Lean and Slippered Pantaloon.
7 The Last Scene.

12. MASONIC AND OTHER LODGE PICTURES.

1 Skull and Cross Bones.
2 Crown and Glory.
3 The Pilgrims.
4 The Knight.
5 Ascension.
6 The Skeleton.
7 Temple of Honor.
8 First Star.
9 Second Star.

10 Third Star.
11 Open Grave.
12 Closed Grave.
13 The Rainbow
14 The Forge.
15 Punishment.
16 Family Happiness.
17 Washington.

13. TAM O'SHANTER.

ILLUSTRATING BURNS' POEM.

1 "And scarcely had he Maggie rallied,
When out the hellish legion sallied."
2 "Gathering her brows like gathering storm,
Nursing her wrath to keep it warm."
3 "The Souter tauld his queerest stories;
The landlord's laugh was ready chorus."
4 "Nae man can tether time or tide;
The hour approaches Tam maun ride."
5 "Ae spring brought off her master hale,
But left behind her ain gray tail."
6 "And vow! Tam saw an unco' sight!
Warlocks and witches in a dance."

14. NEW TALE OF A TUB.

A COMIC POEM ILLUSTRATED.

Each set of these Pictures is accompanied by a copy of the Poem.

1 Opening the Question—the Bengal Tiger.
2 Bengal Ease.
3 The Artful Dodge.
4 Look before you Leap.
5 Under Cover.
6 Increasing the interest of the Tail.
7 The Climax.

15. THE HISTORY AND AMUS-ING ADVENTURES OF REN-ARD, THE SLY FOX.

From the Celebrated Illustrations by Kaulbach, of Munich.

1 Renard leads Bruin to search for honey and entraps him.
2 Renard feigns death, deceives the crows, kills and eats them.
3 Renard deludes the storks, bites off their heads and eats them.
4 Renard entraps the hares.
5 Renard plots the destruction of the chicken family.
6 Renard kills the chickens, but is taken prisoner.
7 Renard condemned to death.
8 Renard taken from prison to cure the king, and succeeds.
9 Renard's cousin, a she-monkey, implores the king to pardon Renard.
10 Renard's life spared, on condition that he fights with the wolf.
11 Renard greases his whole body, so that the wolf cannot lay hold on him.
12 Renard succeeds in blinding the wolf, and conquers him with ease.
13 Renard receives the homage of the courtiers and the people.
14 Renard decorated by the king, and appointed counselor.
15 Renard lives henceforth a life of ease and opulence in his luxurious house.

16. SELECTIONS FROM THE FABLES OF LAFONTAINE.

ILLUSTRATED BY GRANDVILLE.

The Text to these amusing and interesting Fables may be found in E. Wright's English Translation, published by Willis P. Hazard, Philadelphia.

1 The Grasshopper and the Ant.
2 The Raven and the Fox.
3 The Frog that Wished to be as Big as the Ox.

4 The Wolf and the Dog.
5 The City Rat and the Country Rat.
6 The Wolf and the Lamb.
7 The Thieves and the Ass.
8 Death and the Unfortunate.
9 Death and the Woodman.
10 The Fox and the Stork.
11 The Wolf accusing the Fox before the Monkey.
12 The Bird Wounded by an Arrow.
13 The Lion and the Gnat.
14 The Ass Loaded with Sponges, and the Ass Loaded with Salt.
15 The Lion and the Rat.
16 The Dove and the Ant.
17 The Astrologer who stumbled into a Well.
18 The Hare and the Frogs.
19 The Lion and the Ass Hunting.
20 The Miller, his Son, and the Ass.
21 The Wolf turned Shepherd.
22 The Frogs asking a King.
23 The Fox and the Goat.
24 The Drunkard and his Wife.

17. THE CRUSADERS.

1 Harangue to the Crusaders.
2 Departure of the Crusaders.
3 Combat between Crusaders and Saracens.
4 Crusaders besieging a City.
5 Breaking up of the Camp.
6 Return of the Crusaders.

18. THE ILL-FATED SHIP.

1 The Ship Leaving the Wharf.
2 The Ship Sailing with Fair Wind.
3 The Commencement of a Storm.
4 Height of a Storm.
5 The Ship on Fire.
6 The Raft.

19. SCHILLER'S SONG OF THE BELL.

1 Portait of the German Poet, Fr. V. Schiller.
2 The Baptism.

3 Happy Home.
4 Going to Church.
5 Marriage Bell.
6 Harvest Home.
7 Fire Bell.
8 The Tocsin.
9 Funeral Bell.
10 Raising of the Bell.

20. RIP VAN WINKLE.

1 His Scolding Wife.
2 Rip with the Children.
3 Rip at the Village Inn.
4 The Drinking Party in the Mountain.
5 Rip Returns to his Home.
6 Rip Relating his Story.

Class X—Miscellaneous Pictures.

PER SLIDE. $2.50.

1 Alpine Cottage.
2 Approach of Evening.
3 Angels of the Madonna Sixtina.
4 Angel of Peace.
5 Angel of Light.
6 "A Baby was Sleeping."
7 Agriculture.
8 America.
9 Asia.
10 Bacchus and his Panthers.
11 Bargaining for a Horse.
12 Basanio and Portia.
13 Bed Time.
14 Behave Well.
15 Beware.
16 Blessings of the House and of the Field.
17 Bo Peep.
18 Bolton Abbey, in the olden time.
19 Both Puzzled.

"But, sir, if wanst taught be nothin',
then twice taught must be some-
thin', for its doubls what wanst
naught is."

20 Brigand's Hat.
21 Briquet Hound.
22 Broken Doll.
23 Cabman's Leisure Hour.
24 Cavalry Charge.
25 Cendrillon.
26 Charity.
27 Children's Dance.
28 Children's Offering.
29 Children Making Wreaths.
30 Children of Charles the First.
31 Christmas Eve.
32 Christian Maiden.
33 Cocoa Merchant.
34 Come Along.
35 Coming thro' the Rye.
36 Consolation
37 Contentment.
38 Cross of Prayer.

39 Crossing the Thay.
40 Dancing Children.
41 Darwin Expounding his Theory (comic).
42 Daughter of the East.
43 Daughter of Zion.
44 Deceiving Granny.
45 Deer Stalkers.
46 Dinah Consoling Hetty in Prison.
47 Doll's Birthday.
48 Donkey Race (comic).
49 Drawing the Net at Hawesworth.
50 Driving Home the Flocks.
51 Drift Wreck from the Armada.
52 Drumming Lesson.
53 Eagle and Shield.
54 Engineering.
55 Enjoying the Breeze near the Lake.
56 English Cottage.
57 Evangeline.
58 Eve of the Flight.
59 Expectation.
60 Faith.
61 Fairy Tales.
62 Faithful Friend.
63 Family Happiness.
64 Farewell.
65 Farm Yard in Winter.
66 Feeding the Dogs.
67 First Alms.
68 First Born.
69 First Lesson.
70 First Reformer's Protest.
71 Forester's Family.
72 Flute Lesson.
73 Full Practice.
74 Game of Life.
75 Gardener's Daughter.
76 Gathering the Mistletoe.
77 Genius of Electricity.
78 Genius of Steam.
79 Girl Milking.

182 Toilet.
183 The Pets.
184 The Two Dogs.
185 The See-saw.
186 Unconscious Sleeper.
187 Victor of the Glen.
188 Washington Irving and his Literary Friends at Sunnyside.
189 Washing Day.
190 Whitewashing.

191 Winter in New England. ·
192 Winter's Morning.
193 Wounded Hound.
194 Writing Lesson.
195 Youthful Anglers.
196 Youthful Queen.
197 Young Companion and their Hungry Friends.
198 Young Brood.

Class XI—Dissolving Views.

Selected and executed with great care, in sets, so as to produce charming effects in dissolving.

Four Slides, $12.50.

1 Fort Sumter in Time of Peace :
  Daylight,   Moonlight.

Fort Sumter in time of War :
  On Fire during the Bombardment.
  Fire and Smoke Rising.

Three Slides, $10.00.

2 Bay of Naples and Mt. Vesuvius :
  Day,   Night,
  Eruption—Fire and Smoke Rising.

Three Slides, $10.00.

3 Life Near the North Pole
  Day,   Night,
  Moving Aurora Borealis.

Three Slides, $10.00.

4 Castle of St. Angelo and Church of St. Peter at Rome.
  Day,   Night,
  Illuminated and Fireworks.

Three Slides, $10.00.

5 Christmas Eve :
  At Home,   In Camp,
  In Camp, Fire Burning, Smoke Rising.

Four Slides, $10.00.

6 The Four Seasons :
  Spring,   Autumn,
  Summer,   Winter,

Four Slides, $10.00.

7 Voyage of Life :
  Childhood,   Manhood,
  Youth,   Old Age.

Two Slides, $7.50.

8 Water Mill in the Alps :
  Winter,
  Summer—Wheel Moving.

Three Slides, $7.50.

9 The Brave Drummer-Boy and his Father :
  Both Enlist in the Union Army.
  In Battle against the Enemies of the Union.
  Both Mortally Wounded ; they Die together on the Battlefield.

Three Slides, $7.50.

10 Love, Engagement, and Marriage :
  The First Meeting.
  Five Minutes after the engagement.
  Five Years after Marriage.

Three Slides, $7.50.

11 The Hopeful Bride.
  The Happy Mother.
  The Mourning Widow.

Three Slides, $7.50.

12 Courtship for the Second Wife :
  The Ghost of the First Wife Appears.
  The Consternation.

* *Three Slides*, $7.50.

13 Life's Day :
    Morning,   Noon,
    Night.

SETS OF TWO SLIDES.
Per set, $5.00.

14 Cause and Effect :
    Rowing Against the Tide.
    Rowing With the Tide.
15 He who Marries does Well.
    He who does not Marry does
    Better.
16 Morning Star Rising.
    Evening Star Setting.
17 Grace Before Meat.
    Grace After Meat.
18 Expectation,   Satisfaction.
19 Old Woman Reading.
    Old Woman Reeling.
20 Vase of Flowers in Bud.
    Vase of Flowers in Full Bloom.
21 Temptation,   Perdition.
22 Castle of Ehrenfels on the Rhine :
    Summer,   Winter.
23 Conway Castle, England :
    Day,   Moonlight.
24 Windsor Castle :
    Day,   Moonlight.
25 Isola Bella, Italy :
    Day,   Moonlight.
26 The Settlement in the Back-
    woods.
    The First Beginning.
    The Increase.
27 Castle of Chillon on the Lake of
    Geneva, Switzerland :
    Day.
    Moonlight in Winter.
28 Death-bed of the Righteous :
    John Wesley's Last Moments
    Occupied in Praying.
    Death-bed of the Wicked :
    Cardinal Richelieu's Last Mo-
    ments Occupied in Card-
    playing.
29 Abel's Sacrifice Received.
    Cain's Sacrifice Rejected.
30 Noah Building the Ark.
    Noah Receiving Advice from
    Above.
31 Noah's Sacrifice.
    Appearance of the Rainbow.

32 Israelites Passing through the
    Red Sea.
    Destruction of Pharaoh and his
    Host.
33 The Witch of Endor Visited by
    Saul.
    The Witch of Endor raising
    Samuel.
34 Flower Pieces :
    Dahlias and Roses.
    Asters and Poppies.
35 Fruit Pieces :
    Grapes,   Currants.
36 English Bull Dogs.
    English Sheep.
37 The Enemy at His Hiding Place.
    The Enemy at the Door.
38 Rebels Shooting a Prisoner.
    Rebels Defeated.
39 Before the Proclamation.
    After the Proclamation.
40 Vanity,   Junketing.
41 War,   Peace.
42 The Friendly Meal.
    A Temperance Meeting.
43 Joy,   Sorrow.
44 Alexander and Diogenes.
    Jack in Office.
45 The Morning Kiss.
    The Evening Prayer.
46 High Life,   Low Life.
47 Looking In,   Looking Out.
48 Indolence and Folly :
    The Wrong Way of Spending
    the Sabbath.
    The Explanation of the Bible :
    The Right Way of Spending
    the Sabbath.
49 Fondly Gazing (very beautiful).

  " Fondly gazing on that young face,
    With anxious thoughts of future years ;
  The mother watched each budding grace,
  And mused on all her hopes and fears!"

    The Empty Cradle (very beau-
    tiful).

  "Gone! from a world of pain and woe!
    Gone! from death—from sin's alloy,
  Gone! from temptation's wiles and, Oh!
  Gone! gone! from grief to endless joy!

50 Mother's Dream.
    Angel of Peace.
51 Jeff in Power, Surrounded by his
    Generals.
    Jeff Powerless, Surrounded by
    his Captors.

Class XII—Chromatropes, or Artificial Fire-Works, &c.

These slides are singularly curious, the effect being very similar to that of the Kaleidoscope. The pictures are produced by brilliant designs being painted upon two circular glasses, and the glasses being made to rotate in different directions. A pleasing variety of changes in the pattern are caused by turning the wheel—sometimes slowly, then quickly, backward and forward.

1 "The National Flag" Chromatrope. From designs expressly made to introduce the colors of our glorious National Flag. We have five different patterns of this Chromatrope. Each, . $5 00

2 "The Geometrical" Chromatrope. A variety of entirely new and original patterns, of superior Chromatic and Geometrical effects. We have many different styles of this Chromatrope. Each, ... 5 00

3 "The Washington" Chromatrope. A new and beautiful design, with a photographic likeness of Washington in the centre (copied from Stuart's celebrated painting in the Boston Athenaeum), and the stars and stripes revolving around it in glorious array. Each, ... $5 50

4 "The Lincoln" Chromatrope. A correct likeness of our lamented President in the centre of a revolving display of brilliant colors. Each, ... 5 50

5 "The Good Night" Chromatrope. The words "Good Night," encircled by a wreath of flowers in the centre of a Revolving Chromatrope—very appropriate as a closing piece of an exhibition. Each, ... 5 50

6 "Good Night." The words of Good Night in a wreath of flowers—by moving a slip the words appear, ... 3 50

7 Moving Waters. Represents the Waters moving in the Moonlight—a very beautiful and natural effect, ... 2 50

8 Marcy's Eidotrope. 75

PAINTINGS WITH REVOLVING MOTION.

9 View of Old Ruins, which, by being turned around, changes to Portrait of an Old Woman, ... 5 00

10 Views of Rocks and Shrubbery, which by being turned around, changes to Portrait of a Satyr, ... 5 00

11 Holland Windmill, with Revolving Fans, ... 5 00

12 Jeff Davis before the war—a fierce-looking Soldier; by turning the slide, the same painting represents him after the war—a long-eared Jackass, ... 5 00

13 Bombardment of Fort Sumter; the Ironsides throwing Shell, . 5 00

14 Mount Vesuvius—Eruption; throws out Fire and Smoke, . 5 00

15 Snow Slide—representing falling Snow. Each, . . . 4 50

16 The Seven Boys—The Pleiades, ... 5 00

17 Fountain, ... 5 00

18 Newton's Disc, revolving slides, with prismatic colors, for recomposing white light, ... 5 00

19 A Slide representing the rolling up of a Curtain—intended for commencing an exhibition, ... 5 50

This curtain is not needed with the Sciopticon, as its present front arrangement answers the purpose far better, and can be used with any slide.

DISSOLVING CHROMATROPES, FOR USE ONLY IN THE DISSOLVING LANTERNS.

20 "Our Peacemakers," Dissolving Chromatrope. Arranged for dissolving effect, for two lanterns, on two slides. One slider exhibits the National colors in Chromatropic effect, with black centre, for one lantern. The other slider, intended for the other lantern, contains on a movable slider, five life-like portraits of President Lincoln, Major-Generals Grant and Sherman, and Admirals Farragut and Porter, which appear in the centre of the Chromatrope in succession. Per set, ... $12 00

21 "Our Departed Heroes," Dissolving Chromatrope. Arranged for dissolving effect, for two lanterns, on two slides. One slider exhibits the National colors in Chromatic effect, with black centre for the one lantern. The other slider, intended for the other lantern, contains on a movable slider five life-like portraits of distinguished heroes who lost their lives for the preservation of the Union. Per set, . $12 00

☞ Extra Portrait Slides, adapted for the use of the Dissolving Chromatropes. Each slider contains five life-like portraits of distinguished Generals. Per slide, $7.50.

Class XIII—Original Leaf Designs.

Colored and mounted in wool.

PER SLIDE, $1.50.

Motto, "God is Love."
"   "Feed my Lambs."
"   "God will Provide."
"   "Faith, Hope, Charity."
Leaf Cross.

Leaf Anchor.
"   Harp.
"   Cross and Crown.
"   Merry Christmas.
"   Happy New Year.

Class XIV—Statuary and Bas Reliefs.

Backed with opaque paint, and mounted in Wooden Frames.

PER SLIDE, $1.50.

1 Night.   Thorwaldsen.
2 Morning.   "
3 The Four Seasons—Spring.   "
4 The Four Seasons—Summer.   "
5 The Four Seasons—Autumn.   "
6 The Four Seasons—Winter.   "
7 The Council of War.   Rogers.
8 The Fairy's Whisper.   "
9 Taking the Oath.   "
10 Union Refugees.   "
11 The Home Guard.   "
12 The Charity Patient.   "
13 The Returned Volunteer.   "
14 The Wounded Scout.   "
15 The Country Post-Office.   "
16 The School Examination.   "
17 The Picket Guard.   "
18 The Village School Master.   "
19 The Town Pump.   "
20 Mail Day.   "
21 The Bushwhacker.   "
22 Courtship in Sleepy Hollow.   "
23 The Checker Players.   "
24 Uncle Ned's School.   "
25 Apollo Belvidere.

26 The Three Graces.
27 The Greek Slave.
28 Eve, before the fall.
29 Night. } Copeland.
30 Morning. }
31 The Serenade.
32 The Courtship.
33 The Minstrel.
34 Dog Group.
35 The Sentry Box.
36 The Three Companions.
37 The Cymbal Player.
38 The Flower Girl.
39 Flora.
40 Psyche.
41 Clio.
42 Spring Season.
43 Solitude.
44 Woodman's Companion.
45 The Little Market Girl.
46 Industry.
47 Pandora.
48 The Three Companions.
49 Sympathy.
50 The Invalid.

51 The Two Companions.	71 Roman Emperor Caligula.
52 The Flower Girl of Paris.	72 Roman Emperor Caracalla.
53 Courtship.	73 Minerva.
54 Toilet Stand.	74 Rubens.
55 Roman Vase.	75 Michael Angelo.
56 Etruscan Vase.	76 Highland Mary.
57 The Mischievous Brothers.	77 Fisherman's Daughter.
58 Faith.	78 Marguerita.
59 Evening.	79 Jupiter and Hebe.
60 Morning.	80 Sicilian Dancers.
61 Shakspeare.	81 Cupid Asleep.
62 Ariadne and the Tiger.	82 Una and the Lion.
63 Feeding the Pet Dove.	83 The Hunter.
64 Our Saviour—after La Roche.	84 After the Bath.
65 The Little Companion.	85 Amazon attacked by a Lion.
66 Devotion.	86 Amor Indignant.
67 Bedtime Prayer.	87 The Last Drop.
68 Innocence.	88 The Pitcher-Girl.
69 The Pet Dove.	89 The Gladiator.
70 Faith.	

Class XV—Select Painted Comic Slip Slides, giving Laughable Motion to the Figures.

PER SLIDE, $1.25.

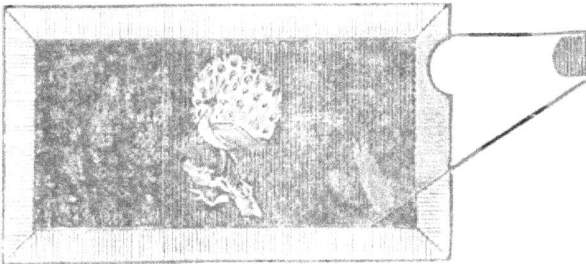

1 Birth of Cupid.	14 French Cook Cooked.
2 Beggar.	15 Good Night, in Wreath of Flow-
3 Boy Smoking.	ers.
4 Boys Birds' Nesting.	16 Jugged Hare.
5 Cottage, with Bridge and Boats.	17 Lady with Expanding Dress and
6 Countryman and Dog Changing	Bonnet.
Heads.	18 Lecture on Tobacco.
7 Cutting Corns.	19 Lion—Moving Eyes and Jaw.
8 Dentist Drawing Teeth.	20 Lovers in Boat.
9 Drinking Fountain.	21 Lodging-house Bedstead.
10 Dog in Kennel.	22 Light-house in Storm.
11 Domestic Shower-Bath.	23 Mischievous Monkey.
12 Diver and Shark.	24 Man Swallowing Rats.
13 Dutchman—Moving Eyes and	25 Monkey Dipping Cat.
Jaw.	26 Magician and Ghost.

3

27 Moving Water and Swan.
28 Man Throwing Stick in Water, and Dog Swimming.
29 Nightmare.
30 Nearing Shore—Dog with Child in Water.
31 Pink Expanding.
32 Pair of Snuffers.
33 Peacock.
34 Performance on Two Chairs.
35 Parrot Pulling off Man's Wig.
36 Rose and Fairy.
37 Somnambulist.
38 Sailor Smoking.

39 Smuggler's Cave.
40 Spider and the Fly.
41 Turk's head—Moving Eyes.
42 Topsy—Moving Eyes.
43 Treading in Father's Shoes.
44 Tax Collector.
45 Taking off Boots.
46 Tulips.
47 Windy Day.
48 Woman with Cat's Head.
49 Woman with Growing Nose and Chin.
50 A Witch.

PER SLIDE, $1.75.

1 Anti-Teetotaler.
2 Acrobats Performing.
3 Barber.
4 Blacksmith.
5 Cupid and Rose.
6 Combing a Bald Head.
7 Cook and Boar's Head.
8 Clown Performing.
9 Death on the Pale Horse.
10 Death in the Cup.
11 Dentistry Improved.
12 Drawing Boot.
13 Don't you Wish you may Get it?
14 Expanding Crinoline.
15 Expanding Carnation.
16 Ferocious Pig.

17 Goose and Tailor.
18 Growing Nose.
19 Good Night.
20 How d'ye Do?
21 Insect Changes.
22 Man Eating Rats.
23 Mischievous Monkey.
24 Nightmare.
25 Pickwick and Widow Kissing.
26 Parson and Punch.
27 Roman Nose.
28 Scene on the Rhine.
29 Stuck Fast.
30 Turk's Head.
31 Tiger's Head.
32 Tailor and Geese.

Class XVI—Paintings with Levers, giving Laughable Motion to the Figures.

PER SLIDE, $2.50.

1 Lady Riding.
2 Stag.
3 Woodman.
4 Moving Chin.

5 Cobbler at Work.
6 Bill Sticker.
7 Man's Face.
8 Horse Drinking.

9 Sambo Lecturing.
10 Stone Breaker.
11 Large Ship by Night.
12 Small Ship and Lighthouse.
13 Boy on Donkey.
14 Swan Drinking.
15 Stag Drinking.
16 Cow Drinking.
17 Beggar.
18 See-Saw.
19 Boy Stealing Apples.
20 Gout.
21 Digger.
22 Monk Praying.
23 Boy Cleaning Boots.
24 Reaper.
25 Fractious Child.
26 Monkey Dipping Cat.
27 Lady Playing Pianoforte.
28 Gent in Pegtops taking off Hat.
29 Horse Eating.
30 Children in Boat.
31 Grooming Horse.

32 Donkey Riding Extraordinary.
33 Sam Weller Cleaning Boots.
34 Woman Beating Boy.
35 Lady Praying.
36 Ship in a Gale.
37 Girl Feeding Goat.
38 Fiddler.
39 Volunteer.
40 Monkey and Fish.
41 Doctor and Patient.
42 Dying Camel.
43 Bathing.
44 Dog of St. Bernard.
45 Native Nursing.
46 Pleasure Boat in Rough Water.
47 Steamboat Pleasure Trip.
48 Goat Feeding.
49 Natives.
50 Look before you Leap.
51 Robinson Crusoe on his Raft.
52 Looking Out for Papa.
53 Signals of Distress.
54 Farmer and Pig.

Class XVII—Dioramic Paintings, with Moving Figures.

On Slides, from twelve to fourteen inches long, with two Glasses, on one of which the scene is painted, and on the other the Figures. The Glass containing the Figures is moved in a groove, and the Figures, Vessels, &c., pass across the Scene.

PER SLIDE, $4.00.

1 Holyrood Chapel.
2 Dover Castle.
3 Bernard Castle.
4 Virginia Water.
5 Conway Castle.
6 Coventry.
7 Lambeth Palace.
8 Sidon.
9 Smuggler's Cave.
10 Newstend Abbey.
11 Pyramids of Egypt.
12 Warwick Castle.
13 Thames Tunnel.

14 Israelites Crossing the Red Sea.
15 Noah Entering the Ark.
16 Menagerie.—Keeper pointing to the cage, in which various animals appear in succession
17 Melrose Abbey.
18 Tintern Abbey.
19 Eddystone Light House.
20 Rome and the Tiber.
21 Arch of Trajan.—Procession of Monks.
22 Nagasaki in Japan.
23 Castle of Chillon.

Class XVIII—Paintings Illustrating Nursery Tales.

1 Old Man, Son and Ass; or, the Folly of Trying to Please Every
   One, eight paintings on two slides, ... $4 00
2 House that Jack Built, ten paintings on two slides, . . . 4 00
3 Jack and the Bean Stalk, eight paintings on two slides, . . 4 00
4 Whittington and Cat, ten paintings on two slides, . . . 4 00
5 John Gilpin, nine paintings on two slides, ... 4 00
6 Cock Robin, ten paintings on two slides, ... 4 00

7 Cinderella, eight paintings on **two slides**, . . .	.	$4 00
8 Robinson Crusoe, eight paintings on **two slides**,	.	4 00
9 Mother Hubbard, eight paintings on **two slides**,	.	4 00
10 Seven Ages of Man, seven paintings on **one slide**,	.	2 00
11 Blue Beard, four paintings on one slide, . .	.	2 00
12 Jack and the Bean Stalk, eight slides, per set, .	.	11 00
13 Puss in Boots, twelve slides, per set, . . .	.	15 00
14 Tale of a Tub, seven slides, per set, . . .	.	10 50
15 Babes in the Wood, per set, ...	.	12 00

Class XIX—Beautifully Colored Photographs of Fine Engravings.

Lists in detail sent on application.

PER SLIDE, $2.50.

The number of Slides in each set.

1 Dore's Illustrations to the Bible, . . .	230
2 Dore's Illustrations to Milton's Paradise **Lost**, . .	50
3 Dore's Illustrations to Dante's Inferno, . .	76
4 Dore's Illustrations to the Legend of the Wandering Jew, .	12
5 Pictures in the Royal Galleries of Dresden, Munich, and Berlin, .	80
6 St. Paul's Journeys Illustrated, ...	34
7 Kaulbach's "Shakespeare Gallery," . . .	12
8 Views on the Bosphorus, ...	80
9 Solomon's Temple, ...	20
10 Selections from Bendemann's Frieze, in Royal Palace, Dresden, .	40
11 Prae-Raphaelite Pictures, by Italian masters, . .	36
12 Scenes Illustrating Cooper's Novels, by Darley, . .	32
13 Commodore Wilkes's Exploring Expedition, . .	40
14 Dr. Kane's Arctic Explorations, . . .	14
15 American Civil War, ...	1000
16 Franco-German War, ...	50
17 Chicago Fire, ...	50
18 The American in Europe, by H. C. Crocket, . .	19
19 Fables of Æsop, ...	24
20 Life and History of a Horse, . . .	8
21 Adventures of a Frog, ...	15
22 Grisetts Grotesques, by Tom Hood, . . .	36
23 Illustrations of Burns's Poem, "Cottar's Saturday Night,"	8

Also many others which, if not in stock, can be made from negatives on hand, at short notice.

SCIENTIFIC DEPARTMENT.

The pictures in the following Scientific *Sets* are photographs beautifully colored, except in such individual cases as do not admit of color.

When these are sealed in Canada Balsam, with carefully finished sky and background, they are sold at $2.50 each. Considering, however, that for educational uses, there is little need for this extra finish, and that their demand for school purposes is likely to be large, the choicest of them have been selected to be sold in sets at an average of $1.50 each. When not taken in sets, they are $2.00 each; when sealed, as above indicated, they are $2.50 each.

These Scientific Slides are just suited to the wants of Educationalists at this time. They need but to be known to be very highly appreciated. These, with the Sciopticon, are far more useful than huge piles of cumbersome philosophical apparatus of many times the cost.

Class XX—Mammalia.

TWENTY SLIDES, PER SET, $30.00.

1 Gorilla.
2 Lion.
3 Tiger.
4 Jaguar.
5 Leopard.
6 Lynx.
7 Hyena.
8 Wolf.
9 Newfoundland Dog.
10 Fox.
11 Grizzly Bear.
12 Seal.
13 Kangaroo.
14 Red Squirrel.
15 Porcupine.
16 Elephant.
17 Rhinoceros.
18 Reindeer
19 Giraffe.
20 Camel.

ADDITIONAL ILLUSTRATIONS,

PER SLIDE, $2.00.

21 King of the Cannibals
22 Skeleton of Man and Gorilla.
23 Mandrill.
24 Diadem Lemur.

25 Vampire Bat.
26 Mole.
27 Hedgehog.
28 Serval.
29 Puma.
30 Shepherd's Dog.
31 Esquimaux Dog.
32 Weasel.
33 Skunk.
34 Raccoon.
35 Brown Bear.
36 Polar Bear.
37 Marbled Seal.
38 Crested Seal.
39 Sea Lion.
40 Walrus.
41 Opossum.
42 Gray Squirrel.
43 Beaver.
44 California Gopher.
45 Bushy-tailed Rat.
46 Brown Rat.
47 Guinea Pig.
48 Red Rabbit.
49 Sloth.
50 Armadillo.
51 Ant Eater.
52 Duck-bill.
53 Skeleton of Elephant.
54 Hippopotamus.

55 Malayan Tapir.
56 Wild Boar.
57 Horse.
58 Zebra.
59 Red Deer.
60 Gazelles.
61 Chamois.
62 Sheep.

63 Cashmere Goat.
64 Musk Ox.
65 Cow.
66 Zebu.
67 Buffalo.
68 Llama.
69 Greenland Whale.
70 Porpoise.

Class XXI—Birds.

TWENTY SLIDES, PER SET, $30.00.

1 Condor.
2 Harpy Eagle.
3 Virginian Eared Owl.
4 Undulated Parrot.
5 Sappho Comet and Crested Humming Bird.
6 Kingfisher.
7 Tailor Bird.
8 Mocking Bird.
9 Skylark.
10 Blackbird.
11 Baltimore Oriole.
12 Blue Jays.
13 Lyre Bird.
14 Royal Bird of Paradise, and Blue Girl.
15 Turkey.
16 Ostrich.
17 White Stork.
18 Flamingo.
19 Domestic Ducks.
20 Pelican.

ADDITIONAL ILLUSTRATIONS,
PER SLIDE, $2.00.

21 Skeleton of a Bird.
22 Turkey Buzzard.
23 Gerfalcon.
24 Imperial Eagle.
25 Bald Eagle.
26 Barn Owl.
27 Cockatoo.

28 Toucan.
29 Mexican Trogon.
30 Parrots.
31 Woodpecker.
32 Black-breasted and King of Humming Birds.
33 Ruby-throated Humming Bird.
34 Whippoorwill.
35 Momotus Ceruleiceps.
36 Missel Thrush.
37 Bell Bird.
38 Blue Bird.
39 Scarlet Tanager.
40 Swallow.
41 Great Northern Shrike.
42 Rose-breasted Grosbeak.
43 Tree Sparrow.
44 Raven.
45 Wild Pigeon.
46 Pheasant.
47 Peacock Pheasant.
48 Ruffed Grouse.
49 Cassowary and Emu.
50 Great Bustard.
51 Whale-headed Stork.
52 Sacred Ibis.
53 Snipe.
54 Curlew.
55 Bean Goose.
56 White Swans.
57 Mutton Albatross.
58 Great Northern Diver.
59 Cormorant.
60 Penguin.

Class XXII—Reptiles and Fishes.

TWENTY SLIDES, PER SET, $30.00

1 Green Turtle.
2 Crocodile.
3 Alligator.
4 Chameleon.
5 Boa Constrictor.

6 Rattlesnake.
7 Viper.
8 Cobra di Capello.
9 Bull Frog.
10 Natterjack.

11 Perch.	16 Flying Fish.
12 Mackerel.	17 Sea Horse.
13 Sword Fish.	18 Sturgeon.
14 Dolphin.	19 White Shark.
15 Fishing Frog.	20 Topedo.

Class XXIII—Insects.

TWENTY SLIDES, PER SET, $30.00.

1 Buprestian Beetles (4 species).
2 Harlequin Beetles.
3 Mole, Cricket, and Cockroach.
4 Katydids, Locusts, and Grasshopper.
5 Chinese Lantern Fly.
6 Bedbug.
7 Dragon Fly.
8 Ant-lion, and Lace-winged Fly,
   with larva.
9 Nymphalis Dissippe, with larva
   and chrysalis.
10 Peacock Butterfly, with larva
   and chrysalis.
11 Papillio Turnus.
12 Parsnip Butterfly, with larva and
   chrysalis.
13 Erebe-strix.
14 Five-spotted Sphinx, with larva
   and chrysalis.
15 Honey Bees—queen worker and
   drone, with comb.
16 Mosquito and Eggs.
17 Proboscis of Horse Fly.
18 Flea (of cat).
19 Human Louse.
20 Walking Stick.

TEN ILLUSTRATIONS ON THE HONEY BEE.

PER SET, $20.00.

1 Queen, Working Bee, Drone, and
   Comb.

2 Head of the Worker.
3 Abdomen of the Worker.
4 Structure of the Eyes of a Bee.
5 Proboscis of the Worker.
6 Wing and Hind Leg of Worker.
7 Sting of Worker.
8 Digestive, Respiratory, and Ner-
   vous System of Bee.
9 Larva and Pupa of Worker.
10 Home of the Bees.

INSECT METAMORPHOSIS.

Showing the different stages of Transfor-
mations, with beautiful landscapes.

PER SLIDE, $2.50.

1 Papillio Machaon.
2 Vannessa Io.
3 Attaeus Lunar.
4 Teigne tapezeila.
5 Vespa Sylvestris.
6 Anthophora Personata.
7 Melontha Vulgaris.
8 Lucanus Cervus.
9 Hydrophilus Piceus.
10 Dystieus Marginalis.
11 Cicindela Campestris.
12 Calandra Palmarum.
13 Phyllium Siccifolium.
14 Locusta Vividissima.
15 Gryllotalpa Vulgaris.
16 Aschna Maculatissima.
17 Cicada Fraxini.
18 Calliphora vomitoria and Sarco-
   phaga Carnaria.
19 Stratiomys Chamæleo.
20 Eristalis tenax.

Class XXIV—Arachnida, Crustacea, et Cœtera.

PER SLIDE, $2.50.

SPIDERS, ARACHNIDA.

1 Lycosa tarentula.
2 Theridion aphane.

3 Epeira diadema, cornuea, angu-
   lata, and bicornis.
4 Scorpion (from Texas).
5 Cheese Mite.
6 Itch acarus.

CRUSTACEA.

1 Bernard Hermit Crab.
2 Lobster.
3 Parthenope Horida.
4 American Edible Crab.
5 Ranine Dentata.
6 Pychnognon Littorale.

ENTOMOSTRACANS.

1 Barnacles.
2 Limulus Longispinus.
3 Daphinia Pulex (male and female).
4 Cyclops Quadricornis.
5 Water Fleas, various kinds.
6 Fairy Shrimp, Chirocephalus.

MOLLUSCA.

1 Paper Nautilus.
2 Octopus or Poulpe.
3 Sepia Officinalis.
4 Pearly Nautilus.

5 Pteroceras Aporrhais and Strombus.
6 Murex Tenuispina and Pyrula canaliculata.
7 Harp Shell.
8 Mitra Episcopalis and Papalis.
9 Cypraea.
10 Turritella, Scalaria, and Vermetus.
11 Helix Albolabris.
12 Pecten Irradians and Mytilus Edulis.

RADIATA.

1 Holothuria, or Sea Cucumber.
2 Sea Urchin.
3 Star Fish.

JELLY FISH, OR ACALEPHS.

1 Pelagia Noctiluca.
2 Cyanea Euplocamia.
3 Physalia Arethusa.

Class XXV—Botany.

VEGETABLE ANATOMY.

TWENTY SLIDES, PER SET, $30.00.

1 Vertical section of extremity of Root (highly magnified).
2 Section of Leaf, White Lily and Oleander (highly magnified).
3 Section of Coniferous Wood, and Glands (highly magnified).
4 Longitudinal section of portion of Stem and Spiral Vessels.
5 Lactiferous vessels of Celandine and Fisus elastica.
6 A Sting of the Nettle, showing circulation of Sap.
7 (1) Air cells from stem Limnocharis Plumieri: (2) ditto, showing open passages at angles of cells: (3) Epidermis of Oncidium altissimum: (4) Stomata of Croton variegatum.
8 Section of Elm Branch.
9 Section of Ash Branch.
10 Transverse and vertical section of Negundo, a year ago.
11 Section of Fern Stem and Scalariform tissue.
12 Polleen Grains (six varieties).
13 Polleen Masses (Orchis, Plantanthera, and Asclepias).
14 Starch Grains (Potato, Wheat, and Maize in cells).
15 Vertical section of Stigma of Ditura.
16 Conducting tissue in Stigma of Ditura.
17 Section of Ovule of Polygonum before and after fecundation.
18 Germination of Fern Spore.
19 Fern and Sporangia.
20 Spores and Sporidia of diseased grain of Wheat.

BOTANICAL ILLUSTRATIONS.

TWENTY SLIDES, PER SET, $30 00.

1 Parts of a plant.
2 Germination.
3 Roots.
4 Buds and Leaves.
5 Flowers and Inflorescence.
6 Stamens and Pistils.
7 Exogenous Structure.
8 Crowfoot Family, Columbine, &c.

9 Pink Family.
10 Tobacco.
11 Clover.
12 Apple.
13 Rose.
14 Melon.

15 Composite Family, Chicory and Calliopsis.
16 Oak.
17 Fir and Hemlock Spruce.
18 Endogenous Structure.
19 Date Palm.
20 White Garden Lily.

Class XXVI—Flowers and Plants.

Skeleton leaves are very beautiful when thrown upon the screen, and even ordinary leaves, petals, &c., show very well.

TWENTY SLIDES, PER SET, $30.00.

1 The White Lily.
2 Lily of the Valley.
3 Holly.
4 Boursalt Rose.
5 Fuschia.
6 Amaryllis Johnsoni.
7 Dahlia Variabilis.
8 Strawberry, Flower and Fruit.
9 Camilla Japonica.
10 Oleander.
11 Magnolia and Passion Flower.
12 Chrysanthemum.
13 Venus' Fly Trap.
14 Peony.
15 Japanese White Lily.
16 Fruit Piece.
17 Nasturtion.
18 Violets.
19 "Consider the Lilies," (with text).
20 White Pond Lily.

PER SLIDE, $2.00.

21 Sarracenia Purpurea.
22 Pelargonium.
23 Almond, Flower and Fruit.
24 Pomegranate.
25 Figs and Olives.
26 Rose.
27 Bunch of Roses.
28 Rudbekia Speciosa, Antirrhinum majus, Lilium Lancifolium.
29 Pine Apple.
30 Scarlet Geranium.
31 Cattleya Superba.

32 Figs.
33 Cactus Triangulaire and Gayac Officinale.
34 Rose and Buds.
35 Citrus Aurantium.
36 Bunch of Fruit.
37 Stock Gillyflower.
38 Blackberry.
39 Passion Flower.
40 Viburnum Opulus and Mespilus Germanica
41 Wellingtonia Gigantea.
42 Cactus (six-sided).
43 Hyacinth.
44 Tulip.
45 The Banyan Tree.
46 Wreath of Flowers.
47 Clover.
48 Tobacco Plant.
49 The Date Palm.
50 Mountain Vegetation of Java.

IMPORTANT PLANTS USEFUL TO MAN.

51 Black Pepper—Piper nigrum.
52 Cinnamon—Laurus cinnamomum.
53 Nutmeg— Myristica moschata.
54 Clove—Caryophyllus aromaticus
55 Coffee—Coffea Arabica.
56 Tea—Thea Bohea and viridis.
57 Cocoa—Theobroma Cacao.
58 Vanilla—Vanilla aromaticus.
59 Opium Poppy—Papaver somniferum.
60 Peruvian Bark—Cinchona cordifolia.

Class XXVII—Physical Geography.

From the best authorities.

PER SLIDE, $2.50.

1 Forms of Snow Crystals.
2 Glacier in Western Norway.
3 Niagara.
4 Great Fall—Yosemite.
5 Canon.
6 Coral Reef.
7 Water Spout.
8 Group of Palms.
9 African Scene.
10 Tropical Vegetation.
11 Banian Tree.
12 Varieties and Distribution of Man.

ILLUSTRATIONS TO HUMBOLDT'S COSMOS.

13 Aurora Borealis.
14 Midnight Sun at the North Cape.

15 Plutonic Rocks, Hartz Mountains, Germany.
16 Plutonic Rocks. Rock labyrinths, near Baden.
17 Primary Rocks. Burning Mountain, near Duttwelles.
18 Primary Rocks. The Lurley Rock, on the Rhine.
19 Volcanic Rocks. Isola della Frizza.
20 Secondary Rocks. The Rock of Gibraltar.
21 Secondary Rocks. Chalk Mountains in Dorsetshire.
22 Secondary Rocks. The Bielgrund, near Dresden.
23 Tertiary Rocks. Tivoli.
24 Alluvial and Diluvial Deposits. The Valley of the Nile.

Class XXVIII—Astronomy.

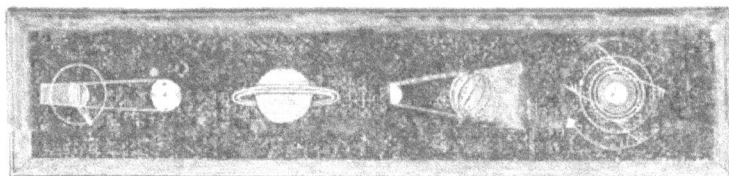

FORTY-ONE PAINTINGS, ON TWELVE LONG SLIDES.

PER SET, $25.00.

Packed in a box, with descriptive book.

1 The Earth's Rotundity (lever movable).
2 New Moon.
3 New Moon. First Quarter.
4 Full Moon.
5 The Moon's Phases.
6 Telescopic View of the Sun.
7 Telescopic View of Mercury.
8 Telescopic View of Venus.
9 The Earth and Moon.

10 Telescopic View of Mars.
11 Telescopic View of Vesta, Juno, Ceres and Pallas.
12 Telescopic View of Jupiter and his Moons.
13 Telescopic View of Saturn and his Moons.
14 Telescopic View of Saturn with rings edgewise and his Moons.
15 Telescopic View of Uranus and his Moons.
16 Orbit of a Comet.
17 Comet of 1819.
18 Comet of 1811.
19 Comet of 1860.
20 Solar System of Ptolemy.
21 Solar System of Copernicus.

22 Solar System of Tycho Brahe.
23 Solar System of Newton.
24, 25 The Sun's Magnitude.
26 Eclipse of the Moon.
27, 28 Eclipse of the Sun.
29 The Moon's Orbit.
30 Different Eclipses of the Moon.
31 The Seasons.
32 The Zodiac.
33 Spring Tide at New Moon.
34 Spring Tide at Full Moon.
35 Neap Tide.
36, 37 Constellation Ursa Major.
38, 39 Constellation Orion.
40 The Milky Way.
41 Nebulæ.

MOVABLE DIAGRAMS.

The motion produced by rack-work.

TEN SLIDES, PER SET, $40.00.

Packed in a box, with lock and key.

1 The Solar System, showing the Revolution of all the Planets, with their Satellites, round the Sun.
2 The Earth's Annual Motion round the Sun, showing the Parallelism of its axis, thus producing the Seasons.
3 The Cause of Spring and Neap Tides, and the Moon's Phases, during its revolution.
4 The Apparent Direct and Retrograde Motion of Venus or Mercury, and also its Stationary appearance.
5 The Earth's Rotundity, proved by a Ship sailing round the Globe, and a line drawn from the eye of an observer placed on an eminence.
6 The Eccentric Revolution of a Comet round the Sun, and the appearance of its Tail at different points of its Orbit.
7 The Diurnal Motion of the Earth, showing the Rising and Setting of the Sun, illustrating the cause of Day and Night, by the Earth's rotation upon the Axis.

8 The Annual Motion of the Earth round the Sun, with the Monthly Lunations of the Moon.
9 The Various Eclipses of the Sun with the Transit of Venus ; the Sun appears as seen through a Telescope.
10 The various Eclipses of the Moon ; the Moon appears as seen through a Telescope.

Illustrating the Moon. Its Topography, Scenery, &c., with a Familiar Descriptive Lecture.

TEN SLIDES, PER SET, $15 00.

1 Map of the Moon. (Beer and Madler.)
2 Diagram illustrating Refraction.
3 The Earth, as seen from the Moon.
4 Telescopic View of the Full Moon
5 Telescopic View of the Moon, first quarter.
6 Telescopic View of the Moon, last quarter.
7 Telescopic View of the Moon, past last octant.
8 Environs of Tycho (from a Photograph by W. de la Rue).
9 Region S. E. of Tycho.
10 View of Copernicus (Naysmith).

TWENTY SLIDES, PER SET, $30.00.

1 Solar System.
2 Phases and Apparent Dimensions of Venus at its extreme and mean distance from the Earth.
3 Inclination of the Axis of the Planets—Venus, Earth, Mars, Jupiter, and Saturn.
4 Diagram illustrating Refraction.
5 Parallels, Meridians, and Zones.
6 True and mean Place of a Planet in its Orbit.
7 Signs of the Zodiac.
8 Telescopic View of the Full Moon
9 Telescopic View of the Moon past the last Quarter.
10 Cause of the Moon's Phases.
11 Mountains on the Moon.
12 Inclination of the Moon's Orbit.

13 Diagram to explain Eclipses.
14 Illustration of the Tides.
15 Telescopic View of Mars.
16 Telescopic View of Jupiter.
17 Telescopic View of Saturn.
18 Comet of 1811.
19 Comparative Size of the Sun and Planets.
20 Star Cluster of Resolvable Nebulæ.

PER SLIDE, $2.00.

21 Bird's-eye View of Saturn and its ring system.
22 Saturn, luminous points visible near the period of the disappearance of the rings.
23 Telescopic View of Saturn.
24 Comparative size of Saturn and the Earth.
25 Telescopic View of the Moon, a little before last quarter.
26 Telescopic View of the Moon, just before the full.
27 Scenery on the Moon.
28 Mountains of the Moon ; view of the region southeast of Tycho.
29 Comparative size of Sun, the Earth and Moon's Orbit.

30 Eclipses and passages of the Satellites of Jupiter, seen from the Earth.
31 Parallax.
32 Direct and retrograde motion of Mercury and Venus.
33 The apparent size of the Sun seen from the principal Planet.
34 Egyptian Zodiac.
35 The Micrometer.
36 Discovery of a small Planet by means of Ecliptic Charts.
37 Ecliptic Chart. From M. Chacornac's "Star Atlas."
38 Herschel's 40 feet Telescope.
39 Measure of the distance of an inaccessible object.
40 Deformation of the Sun's limb at sunset.
41 Convexity of the Ocean.
42 Solar Cyclone, May 5, 1857. (Secchi.)
43 Donati's Comet.
44 Spiral Nebulæ in Virgo. (Rosse.)
45 Nebulæ in Andromeda.
46 Elliptical annular nebula of the Lion. (Herschel.)
47 Spiral Nebulæ in Canes Venatici. (Rosse.)
48 Lunar Crater.

Class XXIX—Geology.

TWENTY SLIDES, PER SET, $30.00.

1 The Geological Record.
2 Ideal Section of the Earth's Crust.
3 Thickness of the Earth's Crust.
4 Section of a Volcano in action.
5 Fingal's Cave.
6 Grotto of Antiparos.
7 Glacier—Mt. Rosa.
8 Glacier Tables.
9 Coral Island.
10 Corals.
11 Rain Drop Marks.
12 Trilobites.
13 Ammonites.
14 Pterichthys—Cocostes, Cephalaspis.
15 Fossil Fern—impression of.
16 A Thrust—in a Coal Mine.

17 Ichthyosaurus.
18 Pterodactyl.
19 Fossil Footmarks.
20 The Mammoth Restored.

PER SLIDE, $2.00.

21 Skeleton of Megatherium.
22 Sigillaria.
23 Lepidodendron.
24 Tracks (the Stone Books).
25 Bone Cavern, Wirksworth, Eng.
26 Skeleton of Hydrarchos Harlanii.
27 Pentacrinites Briareus.
28 Apiocrinites and Actinocrinites.
29 Forest of the Coal Period.
30 Dinornis Mantelli.
31 Foraminifera (from Atlantic soundings).
32 Lava Arch, Iceland.

33 Section of the Cavern of Gailen-reuth (Hartz.)	36 Temple of Serapis (Pozzuolo).
34 Sandstone Columns in Switzer-land.	37 The Dodo (an extinct bird.)
	38 Convoluted Strata.
35 Skull of Mosasaurus.	39 Skeleton of Ichthyosaurus.
	40 Diplacanthus Striatus.

Class XXX—Natural Phenomena.

TWENTY SLIDES, PER SET, $45.00.

1 Rainbow.	11 Drooping Well.
2 Tempest.	12 Coral Reefs.
3 Aurora Borealis.	13 Caverns.
4 Halos.	14 Fingal's Cave.
5 Fata Morgana.	15 Perforated Rocks.
6 Will of the Wisp.	16 Glacier, Mt. Rose.
7 Water Spouts.	17 Glacier Tables.
8 Sand Storm.	18 Icebergs.
9 Geysers.	19 Volcanos.
10 Falls of Niagara.	20 Prairie on Fire.

Class XXXI—Anatomy and Physiology.

TWENTY SLIDES, PER SET, $30.00.

1 Human Skeleton.	10 The Stomach, Liver, and Pan-creas.
2 Human Skull.	
3 Section of the Spine, &c.	11 The Thoracic Duct.
4 Teeth, and structure of same.	12 Heart and Lungs.
5 Muscles, front view.	13 Diagram of Circulation.
6 Muscles, back view	14 Skin and structure of same.
7 Muscles of the head, neck, and face.	15 Brain and Spinal Cord.
	16 General view of the Nerves.
8 General view of the Digestive Organs, in place.	17 Fifth Pair of Nerves.
	18 Facial Nerves.
9 The Digestive Organs.	19 Diagram of the Eye.
	20 Anatomy of the Ear.

A set of 22 slides on Anatomy, copied from 22 plates in Iconographic Encyclopedia (from plate 119 to plate 140 inclusive), on glass three inches square. These, and illustrations of which we have the negatives, will be furnished at $2.50 each. Illustrations of which the negatives must be made to order, will be furnished at $3 each.

Class XXXII—Microscopic Anatomy.

TWENTY SLIDES, PER SET, $30.00.

1 Tessellated and Ciliated Epithe-lial Cells.	3 Longitudinal and transverse sec-tions of Bone, Lacunæ, and Canaliculi, highly magnified.
2 Human Blood Discs, and Blood Discs of Frog.	

4 Muscular Fibres, Fasciculus, and Sarcolemma.
5 Vertical and horizontal section of Stomach, Follicles, and Tubes.
6 (A) Capillary Circulation of Frog's Foot; (B) Capillaries of Air Cells of Human Lungs; (C) Capillaries of Villi of the Jejunum.
7 Origin of Hepatic Veins and Bile Ducts of the Liver Lobules.
8 A Human Malpighian Corpuscle and transverse section of Supra-renal Capsule.
9 Nerve Tubes, Cells, and Ganglia.
10 Transverse section of Human Spinal Cord, close to the third and fourth Cervical Nerves.

11 Pus; (A) from Abscess; (B) Mucus Corpuscles from Schneiderian Membrane; (C) Mucus Corpuscles speckled with Pigment Granules from Larynx.
12 Urinary Deposits; (A) Uric Acid; (B) Oxalate of Lime; (C) Triple Phosphate.
13 Fatty Degeneration of the Liver.
14 Tubercle; (A) in Air Cells of Lungs; (B) Miliary.
15 Scirrhous Growth from Mammary Gland.
16 Tænia Solium.
17 Oxyuris Solium.
18 Trichina Spiralis, mature and in cyst.
19 Liver Fluke, *Distoma hepaticum*.
20 Thrush Fungus, *Oidium albicans*.

Class XXXIII—Optics.

TWENTY SLIDES, PER SET, $30.00.

1 Reflection of Light.
2 Formation of Image by Plane Mirror.
3 Foci of Concave Mirrors.
4 Formation of Image by Concave Mirror.
5 Do. by Convex Mirror.
6 Refraction of Light.
7 Laws of Refraction and total Reflection.
8 Refraction in Body with Parallel Sides.
9 Forms of Lenses.
10 Formation of Image by Convex Lens.

11 Formation of Image by Convex Lenses.
12 Do. by Concave Lens.
13 Spherical Aberration.
14 Action of Prism—Chromatic Dispersion.
15 Chromatic Aberration—Achromatic Prism and Lens.
16 Diagram to explain Wave Lengths
17 Double Refraction—Iceland Spar —Nichol Prism.
18 Polariscope, &c.
19 Colored Rings in Uniaxial Crystals with Polarized Light.
20 Do. in Unannealed Glass Cube.

Class XXXIV—The Microscope and its Revelations.

TWENTY SLIDES, PER SET, $30.00.

1 Tolle's Student's Microscope.
2 Collins' Binocular Microscope.
3 Diagrams. No. 1. Compound Microscope. 2. Huyghenian Eyepiece. 3. Ramsden Eyepiece. 4. Chromatic Aberration.

4 Diagrams. No. 1. Simple Microscope. 2. Spherical Aberration. 3. Diaphragm. 4. Achromatic Objective. 5. High and Low Angle of Aperture.
5 Sheep Tick.
6 Human Head Louse.
7 Dog Flea.
8 Larva of Mosquito.

9 Head of Male Mosquito.
10 Leg of Blow Fly.
11 Eye of Horse Fly.
12 Portion of Wing of House Fly.
13 Scales from Wing of Moth.
14 Wool Fibres.
15 Section of Wheat Straw.
16 Heliopelten.
17 Fine Muslin—showing Cotton Fibre.
18 Foot of Fly—showing Structure of Pads.
19 Proboscis of Fly—Anthromya Lardaria.
20 Saws of Saw Fly.

Additional Photographs of Microscopic Objects enlarged from Nature.

PER SLIDE, $2.00.

21 The Human Louse.
22 The Crab Louse.
23 The Bedbug.
24 The Flea.
25 The Fly.
26 Marine Algae.
27 Hunting Spider.
28 Parasite of Chicken.
29 The Common Mosquito.

30 Eye of Fly.
31 Sting of Bee.
32 Human Itch Insect.
33 Tongue of a Hornet.
34 Tongue of a Bee.
35 Parasite from a Pig.
36 Parasite from a Sparrow.
37 Scale from the Wing of a Butterfly.
38 Parasite of a Beetle.
39 Parasite of a Chaffinch.
40 Parasite of a Field Mouse.
41 Parasite of a Swallow.
42 Parasite of a Mole.
43 Flea of a Mole.
44 Flea of a Mole Pigeon.
45 Eye of a Beetle.
46 Gizzard of a Cricket.
47 Water Beetle.
48 Leaf Insect.
49 Scale from the Wing of a Moth.
50 Saw of the Saw Fly.
51 Spiracle of a Cockchafer.
52 Tongue of a Drone Fly.
53 Trachea of a Silk Worm.
54 Tongue of a Hornet.
55 Transverse Section of Bone.
56 Scale of a Fish.
57 Human Blood Corpuscles
58 Section of a Tooth.

Class XXXV—Crystallography.

Arranged as in Roscoe's Chemistry.

TEN SLIDES, PER SET, $15.00.

1 The Primary Forms of the Six Systems.
2 Secondary Forms of the First or Regular System.
3 Secondary Forms of the Second or Quadratic System.
4 Secondary Forms of the Third or Hexagonal System.
5 Secondary Forms of the Fourth or Rhombic System.
6 Secondary Forms of the Fifth or Monoclinic System.
7 Secondary Forms of the Sixth or Triclinic System.
8 Ice Flowers (Tyndall).
9 Snow Crystals.
10 Ice Crystals.

Class XXXVI—Spectrum Analysis.

TWENTY SLIDES, EACH, $2.75.

1 Decomposition of Light by Prism (Solar Spectrum).
2 Comparative Intensity of Heating Luminous and Chemically Active Rays in Spectrum.
3 Fraunhofer's Map of Solar Spectrum. (1814–15.)
4 The Spectroscope.
5 Spectra of the Sun, Beta Cygni, and Hydrogen.

48

6 Spectra of Potassium, Rubidium, Sodium, and Lithium.
7 Spectra of Carbon Comet II, Brorsen's Comet (1868), Spark and Nebulæ.
8 Spectra of Aldebaran, and Alpha Orionis.
9 Kirchoff's Map (from 194 to 220) and Rutherford's Photograph of same.
10 Spectra of Chlorophyll, Chloride of Uranium, Magenta, and Blood.
11 Gassiot's Spectroscope. Made by Browning.

12 Huggin's Map of Metallic Lines, from 320 to 2790.
13 Huggin's Map of Metallic Lines, from 2790 to 5250.
14 Huggin's Star Spectroscope.

MAP OF SOLAR SPECTRUM.

15 From 38 to 101.
16 From 100 to 163.
17 From 162 to 225.
18 From 224 to 287.
19 From 283 to 345.
20 From 344 to 406.

MAGARGE PAPER MILLS.

ALONG THE WISSAHICKON.—Perhaps no city in the United States has more interesting and romantic surroundings than Philadelphia. Not only is the city itself the very birthplace of American Independence, but its suburbs are all full of historic incidents, and the country round about even yet, in many places, shows the marks of our Revolutionary struggle. The Wissahickon is a stream of most romantic beauty, made doubly attractive by being but an hour's ride distant, either by carriage or street car, from the heart of the busy city.

PLAIN SLIDES

Pictures enumerated in the foregoing lists are beautifully colored by the best artists, and are set in wooden frames. Some fine pictures may have been overlooked in making the selections, and new pictures of surpassing merit, both colored and plain, will doubtless be multiplied as the demand increases. Views of every variety in the market, whether herein enumerated or not, will be furnished at the cheapest rates.

As before intimated, plain lantern transparencies, of varying degrees of merit, are brought out by an increasing number of photographers; a particular catalogue of them will not be attempted, as the lists of to-day might not be such as we would like to stand by to-morrow.

There is a growing demand for plain photographic transparencies from nature, made in the best style of art, and the wish is often expressed that such should be sold at less than a dollar each. It will be seen below that a beginning has been made by marking a class of views of the finest quality at $9 a dozen.

Plain photographs with their protecting glass covers are, for the most part, bound with narrow binding of paper or cloth. They are more liable to be broken than those bound in wood, and more liable to be put in the lantern in a wrong position; but they occupy less space, and, with a grooved slide, they may be passed along, one after another, very conveniently.

Plain foreign views are mostly in the form of half a stereograph, and are sold at $2 each. Any of them will be supplied when called for if they are in this market. Their number is too large to be enumerated here, and such as for their clearness and excellence one would choose for a select list, would be the very ones in most danger of not being in supply when called for.

Class XXXVII—Statuary and Bas Reliefs.

Square pictures in narrow bindings, copied directly from the marbl

PER SLIDE, $1.00; PER DOZEN, $9.00.

1 Apollo, Bust.
2 Ariadne, Bust.
3 Ariadne and the Infant Bacchus.
4 Ariadne on the Sea Dog.
5 At the Bath.
6 Birth of Venus.
7 Blind Homer and his young Companion.
8 Burd Family Monument, St. Stephen's Church, Philadelphia.
9 Breaking Cupid's Bow
10 Cupid, Bust.

11 Cupid Awake.
12 Cupid Asleep.
13 Ceres.
14 Descent from the Cross.
15 Evangeline.
16 Egyptian Girl.
17 Flora.
18 Greek Slave.
19 Greek Slave, Reflected.
20 Gibson's Venus.
21 Goddess of Music.
22 Holy Family.
23 Leda and Swan.
24 Lesbie and the Sparrow.
25 Lighting the Lamp of Silence.
26 Mendicants.
27 Musicians.
28 Our Saviour, Bust.
29 Paul. ⎫
30 Peter. ⎬ Michael **Angelo.**
31 Penelope.
32 Prayer.
33 Prodigal Son.
34 Psyche, Bust.
35 Rebekah at the **Well.**
36 Reading Girl.
37 Red Riding Hood.
38 Rock of Ages.
39 Simply to Thy Cross **I** Cling.
40 Statue of Washington, front **of**
    Independence Hall.
41 The Amazon and the Wild Horse.
42 The Four Seasons—Spring.
43 The Four Seasons—Summer.
44 The Four Seasons—Autumn.
45 The Four Seasons—Winter.
46 The Little Torch **Bearers.**
47 The Wasp.
48 Una and the **Lion.**
49 Venus.
50 Young Student **Reading.**
51 Young Student **Writing.**

Class **XXXVIII**—American Views.

Plain Photographs, three inches square, from nature.

PER SLIDE, $1.00 ; PER DOZEN, $9.00.

PHILADELPHIA.

1 Academy of Music.
2 Horticultural Hall.
3 Independence Hall.
4 New Masonic Temple.
5 New Masonic Hall.
6 Custom House.
7 Exchange.
8 Farmers' Market.
9 Girard College.
10 Ledger Building.
11 Mercantile Library.
12 Post Office.
13 University of Pennsylvania.
14 United States Mint.
15 Beth Eden Church.
16 Cathedral of St. Peter and St.
   Paul.
17 Interior of the Cathedral.
18 Holy Trinity.
19 Jewish Synagogue, Broad Street.
20 Lutheran Church, Franklin St.
21 Methodist Church, Broad and
   Arch Streets.
22 St. Clement's.
23 St. James the Less.
24 St. Stephen's.
25 Continental Hotel.
26 Girard House.
27 La Pierre House.
28 Union League House.
29 Chestnut Street.
30 Chestnut Street Bridge.
31 Market Street.
32 Market Street Bridge.
33 Fairmount Water Works.

FAIRMOUNT PARK VIEWS.

34 Arnold's House.
35 Belmont Mansion.
36 Belmont Glen.
37 Broad Street Entrance.
38 Entrance to Drive.
39 Ferndale Pool.
40 From Old Park.
41 From Mouth of Wissahickon
42 George's Hill.
43 Graff Monument.
44 Grant's Cabin.
45 In the Old Park.

46 Lansdowne Valley.
47 Lincoln Statue.
48 Lemon Hill Mansion.
49 Lemon House.
50 Lover's Walk.
51 North from Basin.
52 North from Sweet Brier
53 North from Laurel Hill.
54 On the Schuylkill.
55 Play Ground.
56 River Scene.
57 Sweet Brier Valley.
58 Summer House.
59 South from the River Road.
60 The Spring.
61 Through the Arch.
62 At Allen's Lane, Wissahickon.
63 At Kitchen's Mill,   "
64 Above Rittenhouse Lane, "
65 Miller's Home,   "
66 Near Hermit's Glen,   "
67 Old Log Cabin,   "
68 Valley Green Bridge,   "
69 Valley Green Hotel,   "
70 Laurel Hill Cemetery.
71 Entrance to Laurel Hill.
72 Old Mortality.
73 Entrance to Mt. Vernon Cemetery.
74 Gardell Monument, Mt. Vernon.
75 Entrance to Woodland.
76 Bailey Monument.

BOSTON.

77 Boston Common.
78 Bunker Hill Monument.
79 Brattle Street Church.
80 Custom House.
81 Faneuil Hall.
82 Fountain Public Gardens.
83 Masonic Hall.
84 Old Elm.
85 Old State House.
86 Public Gardens.

NEW YORK CITY.

87 Broadway.
88 City Hall.
89 Custom House.
90 General Worth's Monument.
91 Opera House.
92 Sub Treasury Building.
93 Washington's Statue, Union Park

BALTIMORE.

94 Battle Monument.
95 St. Peter's Church.

WASHINGTON, D. C.

96 Capitol Building.
97 Interior of the Senate Chamber.
98 Interior of the House of Representatives.
99 Pennsylvania Avenue.
100 Patent Office.
101 Interior of Patent Office.
102 Post Office.
103 Smithsonian Institute.
104 Statue of General Jackson, front of White House.
105 White House.
106 Washington's Statue, front of Capitol.

WISCONSIN.

107 Indian Lodge.
108 Kinnikinnick Falls.
109 Minneanola Falls.
110 Minnehaha Falls, Minnesota.
111 Dalles of St. Croix.

WHITE MOUNTAINS.

112 Mount Washington.
113 View of the Summit House.
114 Tip-top House, summit of Mount Washington.
115 Willey House.
116 Willey Family Furniture.
117 The Basin, Franconia Notch.
118 The Flume,   "   "
119 The Pool,   "   "
120 Walker's Falls, "   "
121 Profile, Old Man, "   "
122 White Mountain Notch.
123 Pulpit Rock.
124 Bridal Veil Falls, White Mountain Vicinity.
125 Frozen Fountain, 30 feet high.
126 Frozen Falls, 300 or 400 feet high.
127 Ice Stalactites.

128 New Post-Office, Portland, Me., White Marble, cost $15,000,-000.

129 Daniel Webster's Place, Franklin, N. H., now used as an Orphan Asylum.
130 Old Mossy Dam, Thornton, N. H.

---

131 Washington's Residence, Mount Vernon.

132 Washington's Tomb, Mount Vernon.
133 Delaware Water Gap, six views.
134 Catawissa Island, four views.
135 Washington's Headquarters, Valley Forge.
136 Ringtown Bridge, Catawissa Railroad.

Class XXXIX—Hymns.

Hymns, &c., will be photographed to order.

PER SLIDE, $1.00; PER DOZEN, $9.00.

1 All Hail the Power of Jesus' Name.
2 Around the Throne of God in Heaven.
3 A Beautiful Land by Faith I See.
4 Beautiful Zion, Built Above.
5 Beautiful River.
6 Beyond the Smiling and the Weeping.
7 Christ our Leader.
8 Climbing up Zion's Hill.
9 Come, Let us Sing of Heaven Above.
10 From Greenland's Icy Mountains.
11 Good Shepherd, Grant Thy Blessing.
12 Hark, the Morning Bells are Ringing.
13 I Have a Father in the Promised Land.
14 I Have a Saviour—He's Pleading in Glory.
15 In the Christian's Home in Glory.
16 I Heard a Voice, the Sweetest Voice.
17 I'm a Pilgrim and I'm a Stranger
18 Jerusalem the Golden.
19 Jerusalem so Bright and Fair.
20 Out on an Ocean all Boundless we Ride.
21 One by One the Sands are Flowing.
22 Over the River.
23 Rock of Ages.
24 Saviour, Like a Shepherd Lead us.
25 Shout the Tidings of Salvation.
26 Shall we Meet Beyond the River.
27 Sweet Hour of Prayer.
28 Sound the Battle-Cry.
29 There is a Happy Land.
30 They are Waiting for Thy Coming.
31 The Golden Shore.
32 There's a Better Way.
33 We are Strong.
34 We are Marching On.
35 Work, for the Night is Coming.
36 Water of Life.
37 While Shepherds Watch their Flocks by Night.

WOODBURY SLIDES.

Class XL—Excelsior Magic Lantern Slides, made by the Woodbury Process.

According to the best judges these views are finer and more beautiful than any plain slides hitherto produced, American or foreign.

Reducing the price of fine lantern pictures from two dollars to seventy-five cents each, opens an easy way for the entrance of the Sciopticon into the home circle with charming effect.

PER SLIDE, $1.00; PER DOZEN, $9.00.

I.—FOREIGN SERIES.

Statuary from the Marbles in Kensington Museum, London.

1 Ino and Bacchus.
2 Queen of the May.
3 Maid and Mischievous Boy.
4 Pandora.
5 The Water Carrier.
6 Justice.
7 Horticultural Gardens, London.
8 Old English Homestead.
9 On the Thames, Richmond.
10 Abergeldie Castle
11 Sir Walter Scott's Tomb.
12 Street in Venice.
13 Grand Canal, Venice.
14 St. Mark's, "
15 Gorge of Pfeffers, Switzerland.
16 Mason Batelier, Ghent.
17 Lion of Lucerne.
18 View in the Grisons, Switzerland
19 Lake of Lucerne.
20 Falkness, Tyrol.
21 The Matterhorn.
22 Kappel Bridge, Lucerne.
23 Interior of St. Peter's Cathedral.
24 East End of Cathedral at Caen.
25 Hotel de Ville, Brussels.
26 The Three Porches at Bayeaux.
27 St. Ouen's Cathedral at Rouen.
28 Panorama.
29 Village of Wessen, Switzerland.

30 Thun, Switzerland.
31 Lake Weggis.
32 Village of Weggis.
33 Cascades, Canton Tessin.
34 Bridge on the Mayenbach Pass of St. Gotbard.

II.—NIAGARA SERIES.

Negatives by Charles Bierstadt.

1 Point View, American and Horse Shoe Falls.
2 Terrapin Tower, winter.
3 Ice Mound and American Falls.
4 Terrapin Tower.
5 American Falls from Pt. View.
6 Horse Shoe Falls and Tower from Goat Island.
7 Ice Bridge, Tower in distance.
8 Winter View from Luna Island.
9 American Falls from "
10 Ice Bridge and Horse Shoe Falls
11 Winter View from Prospect Point
12 Horse Shoe Falls from below.
13 Terrapin Tower from above.
14 Ice Cave under the Banks.
15 American Falls from Luna Isl'd.
16 Hermit's Cascade and first Sister Island Bridge.
17 Horse Shoe Falls from Goat Island, winter.
18 Under the Banks, Niagara.
19 Terrapin Tower from below.
20 Horse Shoe Falls and Rapids.

III.—WASHINGTON SERIES.

1 The Capitol, front view.
2 The Capitol, full view.
3 The Senate, exterior view.
4 House of Representatives.
5 Treasury Buildings.
6 Patent Office.
7 Interior of Patent Office.
8 The Post-Office.
9 The White House.
10 Equestrian Statue Gen. Jackson

IV.—MOUNT VERNON SERIES.

1 The Tomb of Washington.
2 Washington's Mansion.
3 View from Washington's Bath-Room.
4 Ten-Sided Barn built for Washington.
5 Washington Monument, Richmond, Va.

V.—PHILADELPHIA AND VICINITY.

1 Independence Hall, front view.
2 Independence Hall, interior "
3 Independence Bell.
4 Studio in which Gen. Geo. Washington's Portrait was painted.
5 Girard College.
6 Fairmount Waterworks.
7 Lincoln Monument.
8 Union League Building.
9 The Burd Family Monument, St. Stephen's Church.
10 Chew House, Germantown Battle Ground.
11 Washington's Headquarters, Valley Forge.
12 Mill-dam on the Wissahickon.
13 View near Valley Green.
14 Delaware River, instantaneous.

VI.—WATKIN'S GLEN, N. Y.

1 Entrance Amphitheatre
2 Lower Falls.
3 Still-Water Gorge.
4 Central View and Minnehaha.
5 Cavern Cascade and Long Staircase.
6 Mystic Gorge.

7 Glen Mountain House on North Cliff.
8 Cathedral Bridge and Buttermilk Falls.
9 Central Staircase and Mammoth Falls.
10 Central Falls.
11 Looking down Glen Cathedral.
12 Rainbow Falls, from below.
13 Rainbow Falls and Triple Cascades.
14 Frowning Cliff, distant view.
15 Frowning Cliff and Narrow Pass.
16 Gothic Arch of Hope's Studio.
17 Pluto Falls.
18 Artist's Dream.
19 Elfin Gorge and Fairy Pool.

VII.—CALIFORNIA AND OVERLAND SERIES.

Negatives by Charles Bierstadt.

Yosemite Valley.

1 Inspiration Point.
2 Bridal Veil and Three Graces, from Mariposa.
3 El Capitan.
4 "Our Party," Yosemite Valley.
5 Yosemite Falls, 2600 feet high.
6 Lower Yosemite Falls.
7 Mirror View of Yosemite Valley.
8 Eagle Peak.
9 Washington Columns and South Dome.
10 North and South Dome.
11 Mirror View of North and South Dome.
12 Cloud Effect, North and South Dome.
13 Mirror Lake.
14 Liberty Cap.
15 Union Peak.
16 Mirror View of Cathedral Rocks.
17 Grisley Giant, Mariposa Grove.
18 Fallen Monarch.
19 Big Trees, Mariposa Grove.
20 Mariposa.
21 Placer Mining by Chinamen.

Overland Route.

22 Entrance to Devil's Canons, Geyser Springs.

The above cut shows the exact size and shape of the transparency, mat, glass, and the binding of the Woodbury magic lantern slide.

French plain slides, so much valued for their excellence, are of the same width, but are a little shorter.

The disagreeable white sky of most plain slides (relieved somewhat by clouds in the picture above), is now made beautiful on the screen by the blending of colors, effected by a peculiar arrangement lately introduced into the Sciopticon itself.

23 Witches' Cauldron.
24 View on Summit on C.P.R.R.
25 Donner Lake, C.P.R.R.
26 Snow Sheds.
27 Mirror View on Mary Lake, C. P.R.R.
28 Summit on the Sierra Nevadas, 10,000 feet high.
29 New Tabernacle, Salt Lake, Utah
30 Old and New Tabernacles, Salt Lake, Utah.
31 Salt Lake City, Utah.
32 Ogden Canon.
33 Weber Canon, U.P.R.R.
34 Devil's Slide, U.P.R.R.
35 Pulpit Rock, Echo Canon, U.P. R.R.
36 Mormon Fortifications.

VIII.—MISCELLANEOUS SERIES.

1 Old Stone Mill, Newport R. I.
2 Apple Tree in full blossom.
3 Sea Bass weighing 45 pounds.
4 Curious Geological Formation.
5 Collection of Slippers from all Nations.
6 Group of Antiques.
7 Group of Ancient China.
8 Group of Pottery from Peru.
9 Pottery from Herculaneum.
10 Chinese Idols.
11 Mauch Chunk.
12 Delaware Water Gap.
13 Pawnee Indians.
14 Peter La Chere, Chief of Pawnees
15 Prof. Agassiz.
16 Prof. Morse.

Moon—Negatives by L. M. Rutherford.
17 Full Moon.
18 First Quarter.
19 Last Quarter.

Animals and Game.
20 Portrait of Setter Dog.
21 Setter Dog and Snipe.
22 Group of Three Setters.
23 Horse's Head, with fine effect of light and shade.
24 Group of Game.
25 Bunch of Partridges.
26 American Rooster.
27 Meadow View, Media, Pa.
28 A Batrachian Duel.
29 Skeleton Boquet.

30 Sponges, Euplectella **Speoiosa.**
31 Varieties of Corals.
32 Coral Sponge and Star Fish.
33 Coral, Paper Nautilus.
34 Star Fish, Sea Urchins, &c.
35 Conchological **Speoimens.**
36 Sword Fish, Corals, &c.
37 The Hen's Nest.
38 A Basket of Mischief.
39 Stanley dressed as when he met Livingston in Africa.
40 When shall we three meet again?
41 Feed my Lambs.
42 Pulpit Rock, Wisconsin.
43 Kinikinnick Falls, Wisconsin.
44 Highland **Cottage,** Scotland.
45 Suspension Bridge, Niagara.
46 Cattle **Yards,** Chicago.
47 Virtue **in Danger.**
48 Virtue Indignant.
49 First **Parents** Tempted (Darwin)
50 Old Swedes' Church, Philada.
51 American Rooster.
52 Tower **of** London.
53 The Last **Load.**
54 Peace Congress.
55 Council of War
56 Fugitive's Story.
57 Union Refugees.
58 Town Pump.
59 **Charity** Patient.
60 **Dorothea.**

IX.—MICROSCOPIC SUBJECTS.

1 Ant Lion.
2 Texas Spider.
3 Squash Bug.
4 Itch Insect, Human, Female.
5 Young Scorpion.
6 Eye of Horse Fly.
7 Anatomy of the Honey Bee.
8 Anatomy of the Garden Spider.
9 Sheep Tick.
10 Breathing Apparatus of Water Beetle.
11 Proboscis of Blow Fly.
12 The Wicked Flea.
13 Human Louse, Male, Female, and Young.
14 Common Bed Bug.
15 The Fresh-Water Shrimp.
16 American Mosquito, Male.
17 American Mosquito, Female.
18 Blood Corpuscles, Human.

FOREIGN PLAIN SLIDES.

Class XLI.

These slides are now made specially for the lantern, of a greatly improved quality, and in an almost endless variety. We have a fine assortment of these slides, but there is great difficulty in giving a satisfactory catalogue of them. Good *selections* must be made with reference to quality rather than to titles. It requires a long time to replace foreign slides when out of stock, so the catalogue becomes a very faulty guide as to what are at any time obtainable.

The following list has been selected from many thousands, with great care, and ordered in larger quantity in proportion to their beauty and excellence.

PER SLIDE, $1.00.

I.—ENGLAND.

1 Buckingham Palace, London.
2 Ball Room,   Buckingham.
3 Blue Room,   "
4 Drawing Room,   "
5 Gallery of Paintings, "
6 Grand Staircase,   "
7 Promenade Gallery, "
8 Throne Room,   "
9 Charles Street, London.
10 Column of the Duke of York.
11 Greenwich Hospital.
12 Greenwich Observatory.
13 Horse Guards.
14 Mansion House, Lord Mayor's Residence.
15 Panorama of London.
16 Parliament.
17 Parliament and Westminster.
18 St. Paul's Cathedral.
19 Somerset House. ,
20 Tower of London.
21 Trafalgar Square.
22 Waterloo Place.
23 Windsor Castle, London.
24 Dining Room, Windsor.
25 Reception Room,   "
26 Throne Room,   "
27 Castle Street, Liverpool.
28 Docks, Liverpool, instantaneous.
29 Osborne House, Isle of Wight.
30 Fountain of Venus,   "
31 Statue of Queen Victoria, Corridor.
32 Carrisbrook Castle.
33 View of Cowes.
34 View of Ventnor.

II.—SCOTLAND.

1 Balmoral Castle.
2 Panorama of Edinburgh.
3 Monument to Sir Walter Scott.
4 Holyrood Castle.
5 Ruins of Chapel at Holyrood.
6 Melrose Abbey.
7 Cascade at Inversnaid.
8 Gorge at   "
9 Lake Katrine.
10 Rumbling Bridge at Dunkeld.
11 Trossack's Hotel and Benevenue

III.—FRANCE.

1 Panorama of Paris. 8 Bridges.
2 Perspective of Bridges, from the Luxembourg side.

3 Perspective of Bridges, from the Tuilleries side.
4 Port St. Nicholas.
5 Colonnade of the Louvre.
6 Pavillion of the Clock, "
7 Assyrian Room, Museum of the Louvre.
8 Egyptian Room.
9 Gallery of Paintings, Museum of the Louvre.
10 Statue of Diana, Museum of the Louvre.
11 Rameses, Museum of the Louvre
12 Three Graces, "   "   "
13 Venus de Milo," "   "   "
14 Cafe Chantant, Champs Elysees.
15 Facade of the Grand Hotel.
16 Hotel de Ville.
17 Museum de Cluny.
18 New Opera House.
19 Place de la Concorde.
20 Rue de Rivoli.
21 St. James' Tower.
22 Triumphal Arch du Carousel.
23 The Door   "   "
24 Triumphal Arch de l'Etoile.
25 Fountain Catherine de Medicis.
26   "   Cuvier.
27   "   of the Innocents.
28 Palace of the Corps Legislatif.
29   "   Instituto.
30   "   Industry.
31   "   Justice.
32   "   Luxembourg.
33 Throne Room, Luxembourg.
34 Royal Palace.
35 Tuilleries "
36 Room of Apollo, Tuilleries.
37 Gallery of Diana,   "
38 Throne Room,   "
39 Church St. Augustin.
40   "   "   interior.
41 Holy Chapel.
42 Church of the Invalids.
43   "   "   interior.
44 Chapel of Tomb of Napoleon I.
45 St. Germain l'Auxerrois.
46   "   "   interior.
47 The Madeleine.
48 St. Vincent de Paul.
49 Trinity Church.
50 Lake and Castle of Pierrefonds.
51 Castle Maintenon.
52 Chateau de Fontainbleau.

53 Chateau de Chambord.
54 Palace of Versailles.
55 Battle Gallery, "
56 Glass   "   "
57 Great Trianon.
58 Gallery of Fine Arts, Trianon.
59 Bed of the Empress Josephine.
60 Room of the Crusades.
61 Panorama of Rouen.
62 Notre Dame, Rouen.
63 Cathedral of Tours.
64 St. Ouen.
65   "   interior.
66 Panorama Strasburg.
67 Lyons, Perspective of the Saone
68 Hotel de Ville, Lyons.
69 Marseilles.
70 Palace of the Bourse, Marseilles
71 General View of Havre.
72 Chateau Pau, Pyrennes.
73 Convent of the Grand Chartreuse
74 Hotel   "   "   "

IV.—SPAIN.

1 Panorama of Madrid.
2 Palace Royal,   "
3 Fountain of Seasons, au Prado.
4 Queen's Room, Escurial.
5 Court of Lions, Alhambra.
6 Fountain of Lions,   "
7 Colonade of Pilate's House, Seville.
8 Fountain, Pilate's House, Seville
9 Palace of the Duke de Montpensier.
10 Panorama of Cadiz.
11 Gallery of Statues, Casa del Labrador.

V.—ITALY.

1 Panorama of Genoa.
2 Cathedral of Sienne.
3   "   "   interior.
4 Florence.
5 Palace Pitti.
6 Tomb of Laurent de Medicis, by Michael Angelo.
7 Venus de Canova, by M. Angelo.
8 Statue of Dante and Church of the Holy Cross.
9 Panorama of Pisa.
10 Baptistery   "
11 Campo Santo, "
12   "   "   "   interior.

13 Leaning Tower, Pisa.
14 Panorama of Venice.
15 Arsenal, Venice.
16 Bridge of Sighs.
17 Ducal Palace, Venice.
18 Giant's Staircase, side.
19 " " front.
20 La Logia.
21 Palace of the Foscari.
22 " Vendramin.
23 St. Mark's.
24 Facade, Church of St. Mark s.
25 Street in Venice.
26 Rialto.
27 Panorama of Rome.
28 Arch of Constantine.
29 Arch of Septimus Severius.
30 Arch of Titus.
31 The Coliseum.
32 The Forum.
33 The Forum of Trajan.
34 The Tiberine Isle.
35 Chateau and Bridge St. Angelo.
36 St. Peter's Church.
37 " " interior.
38 Genius of Death, by Canova.
39 Statue of Moses, by M. Angelo.
40 Library of the Vatican, interior.
41 Statue of Venus, Vatican.
42 Gallery of the Bras Neuf.
43 Statue of Diana, Vatican.
44 " Fortune, "
45 Statue of Apollo Belvidere.
46 Perseus, by Canova, Belvidere.
47 Group of the Laocoon, "
48 Statue of Apollo, Capitol.
49 Gallery of the Palace Colona.
50 Statue of Venus, Villa Borghese.
51 Room of Antique Marbles, Villa
     Borghese.
52 St. Paul's Church, interior.
53 Cascade of Terni.
54 Naples.
55 Port of Pozzuoli.
56 Temple of Serapis, Pozzuoli.
57 Perspective of the Basilic Pom-
58 House of the Fawn.   [peii.
59 " " Musicians.
60 " Cornelius Rufo.
61 Questor.
62 The Way of the Tombs.
63 Triumphal Arch and Forum.
64 Villa Diomede.
65 Amphitheatre.

VI.—SWITZERLAND.

1 Geneva and Lake.
2 Village of Chamouni.
3 Cascade of the Arve.
4 Crevasse Mer de Glace.
5 Dangerous Passage of the Bossons
6 Needles of the Bossons.
7 Ascent of Mt. Blanc.
8 Grand Mulets, Dome of Gouter.
9 Mt. Cervin.
10 Port of Lucerne.
11 The Simplon.
12 Valley of Gondo Simplon.
13 The Muotta, Switz.
14 Grindelwald.
15 Glacier Grindelwald.
16 Lake and Hospital, Grimsel Pass
17 Falls of the Staubach.
18 Hotel and Castle, St. Gervais.
19 Berne.
20 Zurich.
21 Fribourg.
22 Castle of Chillon, Lake Geneva.
23 St. Bernard Hospital.
24 Bridge on the Albule Grisons.
25 Ortenstein Castle, "
26 Devil's Bridge, "

VII.—GERMANY.

1 Palace Royal, Berlin.
2 Gallery of Portraits, Berlin.
3 Throne Room.
4 The Library.
5 White Chamber.
6 Palace of the Prince Royal.
7 Interior of the Museum, Berlin.
8 The Amazon.
9 French Church, Berlin.
10 Royal Theatre, "
11 Unter Der Linden. "
12 Cologne, Rhine.
13 Cathedral, Cologne.
14 General View of the Cathedral.
15 St. Martin's Church "
16 Cathedral at Bonn.   [lentz.
17 Ehrenbreitstein and Bridge Cob-
18 Heidelberg.
19 Mayence.
20 Cathedral, Mayence.
21 Oberwessel.
22 Pfatz, Caub, Gutenfels.
23 Rheinstein.
24 Rheinfels.

Class XLII—Effect Slides.

These slides are new, beautiful, large, and fine ; having been recently modernized from old designs. They, of course, require a pair of dissolving lanterns.

1 St. Goar on the Rhine.—Landscape at night. Clouds passing. The moon sometimes appears and illuminates the landscape and buildings. Two slides. $10.50

2 Icebergs in the Northern Sea.—The ice is seen moved by the tide, a ship is seen being wrecked by the iceberg. Two slides, $7.00

3 Ship at Sea.—The ship is seen under full sail, the weather becomes stormy, and the ship wrecks on a cliff under repeated lightning. The sky clears. The wreck is seen in the background. The crew are saved in the boats. A rainbow appears gradually in the heavens. Five slides. $15.00

4 Interior of the Cathedral at Cologne, by day.—Night comes on gradually, and the church is seen filling with people. Two slides, $15.00

5 Jerusalem, with the Cross of Christ in the foreground.—The cross remains and the picture gradually changes to the interior of the Church of the Holy Sepulchre. Then clouds form around the cross, in which angels appear. Three slides, $18.00

6 St. Peter's Church at Rome, by day.—Night comes on. The church is seen by moonlight, and is then illuminated. Three slides, $13.50

7 Summer Landscape in Switzerland.—A cottage in the foreground takes fire and burns to the ground. The ruins are now seen by moonlight. Three slides, $14.00

8 A Mill in England.—The water-wheel is seen in motion, and the flowing water beneath. Winter sets in, and the wheel appears frozen fast. Two slides, $12.50

9 Melrose Abbey, Scotland. The rising moon is seen through one of the windows, $5.50

10 Stockholm by Moonlight.—A steamer has just arrived, from which smoke ascends. The moon glistens on the water. A boat with passengers is seen approaching the landing. Three slides, $9.00

11 Porch of the Cathedral at Muuster, by Moonlight.—The window is illuminated, the door gradually opens, the interior is seen filled with people, and High Mass is being performed. Two slides, $8.00

12 The Rocks of Heligoland, by day and night.—Travelers light a fire, the light from which is seen glimmering on the most elevated rocks, with a very beautiful effect. Three slides, $11.50

13 Trenton Falls, New York.—The water is seen pouring over the Falls and flowing away; this has a most beautiful effect. Two slides, $8.00

14 Bombardment of Sebastopol.— A frigate steams up and opens the attack by firing a Columbiad. The fort returns the fire, and the bombs are seen flying through the air. Three slides, $10.00

15 St. Petersburg by Moonlight.— The moon rises and the glistening on the water is very beautifully shown. Two slides, $7.00

16 Interior of Dining Hall, Holyrood Palace.—Knights are seen seated at the table drinking. One of the knights has risen and is looking out of the window. The moon is seen rising, and the interior is seen by moonlight. Two slides, $7.00

17 Castle on Lago Maggiore.—A lover approaches in a boat to serenade, upon which a lady appears on the balcony. One slide, $4.50

18 Church of St. Nicholas, Hamburg, before the great conflagration.—The church is then seen on fire and is burnt to the ground; the ruins are seen by moonlight, and at last the new church is seen entirely finished. Four slides, $20.00

19 Tower of London, Day, Night, and on Fire.—Three slides, $10.50

20 Views in the Arctic Regions.— Night sets in, and the Aurora Borealis is seen in the heavens. Three slides, $9.00

21 Mosque of Omar, Day, Night, and Illuminated.—Three slides. $5.50

22 Water Mill in Pennsylvania.— The wheel is seen in motion; a swan swims on the water and drinks; the moon rises and glistens on the water; the windows in the mill are now lit up; the ground is seen covered with snow, and the snow is seen falling. Five slides, $20.00

23 Falls of Niagara from the Canada side, in the mist below, a beautiful rainbow gradually appears. Two slides, $6.00

24 Destruction of Moscow—Panoramic View of Moscow.—Twilight sets in. Moon appears. View changes to winter. Ground is covered with snow, and snow is seen falling. Fire breaks out near centre of city, and gradually spreads until the entire city is consumed. Five slides, $25.00

25 Destruction of the First Born.— View of an Egyptian City. Angels appear in the sky and cast thunderbolts over the heavens. Two slides, $6.00

26 Salisbury Cathedral.—View of the Cathedral, the same illuminated at night. Changing into a moonlight view. Three slides, $8.00

27 Faust and Marguerite.—Faust as seen in his laboratory. Mephistophiles with magic bowl

in his hand. Suddenly the vision of Marguerte appears, and flames dart out from the bowl. Two slides, $8.00

28 Star of Bethlehem.—Shepherds are seen seated on the ground, and in the distance the star appears and gradually approaches until the Son of Man is seen in a halo of glory. Three slides, $8.00

29 Westminster Abbey by Daylight. —Scene changes to moonlight, and interior is seen lighted up. Three slides, $8.00

30 Magician and Cauldron.—Cave, and magician with his magic wand—with cauldron in corner, out of which appears at various times, ghosts, witches, hobgoblins, etc., etc. Two slides, $9.00

31 Napoleon.—Powerful at the head of his army. Scene changes to Napoleon powerless on the barren rock at St. Helena. Two slides, $8.00

32 White Chamber in the Royal Palace, Berlin, with the Ghost. —The best effect yet produced. Two slides, $5.00

33 Mercy's Dream.—Mercy is represented in a reclining position beneath a spreading tree. An angel from Heaven appears, and places a crown of glory on her head. Two slides, $5.00

34 Angel of Peace.—A beautiful landscape showing a city at night, with the new moon in the sky reflected in the water. The figure of an angel bearing a child appears like a vision in the sky, and then fades away. Two slides, $5.00

35 The Rock of Ages.—A dark and stormy sky, and the waves dashing against a stone cross (the Rock of Ages), are here represented. A wreck is seen in the distance. The wreck disappears, and the figure of a woman appears clinging to the cross. Two slides, $5.00

36 Train of Cars.—A railroad bridge in a dark forest is seen by moonlight. A train of cars dashes by, the headlight and sparks flying from the engine making a very brilliant appearance. Two slides, with movement, $6.00

LANTERN READINGS.

We are now prepared to furnish Descriptive Lectures to accompany sets of slides of the following subjects:

1 Egypt and the Nile.
2 Syria and Palestine.
3 Life and Travels of St. Paul.
4 Pilgrim's Progress.
5 Christiana and her Children.
6 Rome, the Capitol of the Caesars and the Popes.
7 Tour in Switzerland.
8 Tour on the Rhine.
9 Paris.
10 Ireland.

11 China and the Chinese.
12 Astronomy.
13 Natural Phenomena.
14 The Tabernacle and Temple.

In preparation.

15 The Cities of S. E. Europe.
16 The Cities of N. E. Europe.
17 Italy.
18 American Scenery in the Far West, etc.

INDEX.